Games, Sex and Evolution

By the same author

The Theory of Evolution (Penguin Books, 1956)

The Evolution of Sex (Cambridge University Press, 1978)

Evolution and the Theory of Games (Cambridge University Press, 1982)

The Problems of Biology (Oxford University Press, 1986)

Games, Sex and Evolution

John Maynard Smith

Professor of Biology
University of Sussex

HARVESTER·WHEATSHEAF

NEW YORK   LONDON   TORONTO   SYDNEY   TOKYO

First published 1988 by
Harvester · Wheatsheaf
66 Wood Lane End, Hemel Hempstead
Hertfordshire HP2 4RG
A division of
Simon & Schuster International Group

Printed and bound in Great Britain by
Billing & Sons Ltd, Worcester

British Library Cataloguing in Publication Data

Smith, John Maynard
Games, sex and evolution.
1. Evolution — Mathematical models
2. Game theory
I. Title
575.01   QH371
ISBN 0–7108–1216–7

1 2 3 4 5   92 91 90 89 88

Contents

Part 1   Science, Ideology and Myth   1

1. How to win the Nobel Prize   3
2. Storming the Fortress   8
3. Symbolism and Chance   15
4. Science and the Media   22
5. Molecules are not Enough   30
6. Science, Ideology and Myth   39

Part 2   On Human Nature   51

7. The Birth of Sociobiology   53
8. Models of Cultural and Genetic Change (written with N. Warren)   61
9. Constraints on Human Behaviour   81
10. Biology and the Behaviour of Man   86
11. Tinkering   93
12. Boy or Girl   98
13. Genes and Memes   105
14. Natural Selection of Culture?   114

Part 3   Did Darwin get it Right?   123

15. Palaeontology at the High Table   125
16. Current Controversies in Evolutionary Biology   131
17. Did Darwin get it Right?   148
18. Do we need a new Evolutionary Paradigm?   157

Part 4   Games, Sex and Evolution   163

19. Why Sex?   165
20. The Limitations of Evolution Theory   180
21. The Evolution of Animal Intelligence   192
22. Evolution and the Theory of Games   202

Part 5   The Laws of the Game   217

23. The Counting Problem   219
24. Understanding Science   225
25. Matchsticks, Brains and Curtain Rings   231
26. Hypercycles and the Origin of Life   237
27. Popper's World   244
28. Rottenness is All   250

Acknowledgements   258
Index   261

PART 1

SCIENCE, IDEOLOGY AND MYTH

I have always been fascinated by the history of science. *How to Win the Nobel Prize*, and *Storming the Fortress*, are reviews of books about this history. Both are written by justly famous scientists who happened to be professors at Harvard, but they have little else in common. Watson's book is a highly personal account of one crucial discovery; Mayr's is a history of the whole of biological thought, but in its own way it is as personal a book as Watson's. *Symbolism and Chance* is my own attempt at the history of ideas, but it is no more than a sketch of a history I would like to write. Unfortunately, one has to choose between describing science, and doing it. This same conflict comes up in a different form in *Science and the Media*. This was delivered as the presidential address to the zoology section of the British Association for the Advancement of Science in 1983. The science correspondents of all the quality papers were there. Although the meeting was extensively reported, my address was not mentioned once. Apparently it is not news to *The Times* to be told that their record has been miserable. I should have known that one cannot win an argument with the press.

Molecules are not Enough masquerades as a review of a

1

2

Science, Ideology and MythScience, Ideology and Myth

book by Levins and Lewontin. Really it is an attempt to come to terms with my own past. Levins is a scientist who has always been a Marxist, and Lewontin one who has become a Marxist: I am a scientist who has ceased to be a Marxist. The essay should be read with that in mind.

Science, Ideology and Myth was originally delivered as the Bernal lecture to the Royal Society in 1980. The Society decided that it was not suitable for publication in any of their journals, but versions of it have since been published in the US by Natural History, and in France by La Recherche. It overlaps somewhat with *Symbolism and Chance*: the two essays were written at much the same time, when I was trying, rather unsuccessfully, to discover whether there was any sense in structuralist anthropology. It deals with topics that have worried me ever since I was at school. What is the nature of scientific theories, and how do they differ from religious beliefs and political convictions?

1

How to win the Nobel Prize

Will Jim remember what he learnt at the seminar at King's by the time the train has carried him back to Cambridge? Will Francis stay at the wine-tasting party, or will he go home and write down the crucial equation? Will Bragg insist on Crick finishing his PhD thesis and so stop him from working on DNA? (By the way, did that boy ever get his PhD? I finished the book worried that he might never make it.) How can the Cambridge group get a good look at the King's X-ray photographs? Will Jim foil Rosy Franklin and win his Nobel Prize? Read the next thrilling instalment and learn the answers!

Professor James Watson's account of how the structure of the material of which genes are made came to be discovered has all the compulsive qualities of the best whodunits or science fiction stories. First, it is a very funny book: the author never misses the opportunity to tell a good story, even when the joke is against himself. There is the account of how Chargaff, when a guest at a Trinity College dinner, dismisses Watson and Crick as unworthy of serious consideration because Crick cannot remember the formulae of the nucleotides. Better still is the account of the visit to Cambridge by Maurice Wilkins and Rosalind Franklin to see the first model of DNA built by Crick and Watson. At first all goes well, but then it becomes apparent that the model is based on a misrecollection by Watson of the water content of the samples of which the X-ray photographs had been taken. The scene is described with an economy and precision of which any novelist could be proud. More important, the book conveys the excitement associated with scientific discovery. As the search nears its goal, and rumours reach Cambridge from across the Atlantic that the old maestro Pauling is hard at it, the tension reaches a climax. Most important of all, this book has as its central theme, not an imaginary murder or a suspension of the laws of physics, but a genuine intellectual puzzle whose solution the reader can follow as it emerges.

It may seem odd to start discussing a book which describes the most important discovery in biology since Darwin, and is written by one of the main protagonists, as if it were a fictional adventure story. There are reasons for starting in this way, which I hope to make clear. But what is one to say of the book as a contribution to the history of science? First, we must be clear what is it the author has tried to do. He has not set out to write an objective account of what happened, but to tell us what it seemed like to him at the time. This is an entirely laudable intention. I would be delighted to have an equally frank and subjective account of their discoveries by, for example, Gregor Mendel or Isaac Newton.

In many ways, Professor Watson is admirably equipped to write such an account. He has an eye for the significant detail which will bring a scene alive. No sense of false modesty prevents him from saying what he thinks of other scientists. He points out – let us face it, not without truth – that most scientists are stupid, and that most of them are working on trivial problems. He is no hypocrite; indeed it has clearly never occurred to him that it might be proper to pay a tribute to virtue. Consequently, he is prepared to allow the motives for his own actions to emerge with great clarity. If he is sometimes uncharitable in the motives he ascribes to others, at least he never pretends that he is holier than they.

Granted, then, that this is intended to be an account of what it felt like to him at the time, how accurate an account is it? I have little doubt that the picture which emerges of the author and of the part he played is reliable. But I suspect that those who accept his account at its face value will be seriously misled. There are two reasons for this. First, to give an accurate account of one's part in great events it is not sufficient to be frank; it is also necessary to be self-critical. The author remarks that he has never met Francis Crick in a modest mood. I have never met Professor Watson; if I do, I do not expect to find him in a self-critical mood.

The other reason why this account must be taken with a grain of salt is that the author is a natural raconteur. I like telling stories myself and I recognise the technique. The best stories are true stories which have been bent a little to meet the requirements of art; one adds corroborative detail to bring verisimilitude to an otherwise bald and unconvincing narrative. In trivial matters, Professor Watson's habit of embroidery helps the story to go with a swing. Francis Crick's laugh is not quite as piercing as he would have us believe, but it is fun to imagine it that way. It is more serious when it leads him to paint a portrait of Rosalind Franklin which is perhaps the most unsympathetic and unattractive account ever given by one scientist of another. As far as one can discover, his only direct grounds for complaint were that in a handful of meetings Rosalind Franklin – wrongly, as it turned out – was sceptical about the value of model-building as a means of elucidating the structure of DNA. In the author's defence, it could be said that if that is the way he felt about her at the time it is honest of him to admit it, and that he makes handsome amends in the epilogue.

What we have, then, is a book which is enormous fun to read, and which, provided it is taken less as an account of what Professor Watson felt like at the time than as an account of what he would like us to think he felt at the time, tells us a very great deal about how the structure of DNA was discovered. In most of its features, the picture which emerges need not surprise anyone who has thought about the nature of scientific discovery. There is no planned,

logical and inexorable attack on the problem. Instead there
are false starts, insights which turn out to be wrong, and
periods when the problem is dropped altogether because
there is no obvious way forwards. But if fundamental
scientific discoveries could be made by the application of
logic and hard work, science would be a dull business, and
not worth writing books about or giving Nobel prizes for.
Those – they must be fewer today than they were in my
childhood – brought up with a picture of the scientist
working selflessly in the service of humanity will be shocked
by the picture of competitiveness between individuals and
laboratories which emerges. They should not be. To mind
about priority is nothing new: Newton and Leibnitz
quarrelled about who had invented the calculus. More
immediately, how would you feel if you were an unknown
beginner on the brink of a discovery which would put your
name in the history books, and you knew that one of the
world's most distinguished chemists was on the same trail?

To me, as a university teacher, perhaps the most significant
fact to emerge from this story is how utterly unprepared
Watson was in terms of formal education to make this
particular discovery. When he went to Cambridge, he knew
virtually no biochemistry and was obstinately determined to
learn none, and he knew nothing of X-ray crystallography.
What he had was a conviction that the structure of DNA
was the most important problem in biology, the courage (or
arrogance – in intellectual matters they come to much the
same thing) to hope that he could solve it, and the intuition
that the way to solve it was by model-building based on X-
ray photographs. He also needed some luck: to find Crick
as a collaborator calls for good luck as well as good
judgement.

Today, fifteen years after the publication of the
Watson–Crick paper on the structure of DNA, it is becoming
fashionable to sound the alarm about molecular biology.
Soon, it is argued, it will be possible so to interfere with
human reproduction that we shall destroy family life, or to
develop more virulent cancer-inducing viruses which could
decimate mankind: the time has come therefore to call a
halt. I am inclined to agree that if you want to kill very

large numbers of people, it will soon be cheaper to do so by applying molecular biology than atomic physics. If so, there are clearly grounds for being worried about the applications of molecular biology. But there are no grounds for being silly about it. One way of being silly is to suggest that those engaged in fundamental research should at the same time worry about the possible applications of their as yet unmade discoveries. This suggestion would make some sense if the scientist were presented with a little box containing the secret of DNA. He could then devote his energies to thinking about the possible consequences of opening the box. But life is not like that. Scientific discovery is difficult. No one who reads Professor Watson's book will regard it as a sensible suggestion that he ought to have been thinking about the social consequences of his discovery, in the unlikely event of his getting there before Pauling. If he had any thoughts to spare, it is not surprising that they were more often devoted to au pair girls than to the social consequences of science.

It would be possible to arrest the advance of fundamental scientific research if governments refused to finance it, and if we persecuted those who indulged in it. In so doing, we should be ending the most successful attempt yet on the part of mankind to satisfy their curiosity about themselves and the world they live in. Fortunately, there is no need for such drastic measures if what we are concerned about are the applications of molecular biology. The atomic bomb did not design and drop itself, and in the same way more virulent pathogens or more potent toxins will not happen of themselves. If you are worried about the applications of molecular biology, the sensible thing is to campaign for a government which will close down Porton and give the money they save to the Medical Research Council.

No doubt a more objective account of the discovery of DNA will be written. For sheer excitement, it is unlikely to equal Professor Watson's book. We must be grateful to him on two counts: first as an entertainer in a dull world; and second, for telling us what it is like to be in at the kill.

2

Storming the Fortress

To write a history of the world, a man would have to be slightly crazy. He would have to be possessed by some single idea which, he believed, could illuminate the whole course of history, and in the light of which all events could be judged. To write a history of biological thought, particularly one confined to thoughts about biological diversity rather than physiological mechanism, is a less ambitious undertaking, and Ernst Mayr is certainly not crazy. But the task is sufficiently formidable, and Mayr has succeeded at it only because he does indeed have single, unified vision, which he uses to evaluate scholars from Aristotle to Theodosius Dobzhansky and R. A. Fisher.

His vision is that the history of biology has been a struggle between two world views – essentialism and population thinking. The essentialist view holds that there are a finite number of kinds of animals and plants, each kind characterised by certain essential features which it is the business of the biologist to recognise. These 'kinds' correspond to what we today call species. The members of a species share an identical essence. They may differ from one another in various ways, as people differ in height, hair colour and fingerprints, but these differences are accidental and unimportant; they do not alter the essence. The populational view holds that individuals are bound together, not by the possession of a common essence, but by the fact that they interbreed.

The supreme essentialist was Linnaeus. His attitude derived in part from certain features of the real world (which I shall describe in a moment) and in part from the philosophy of Aquinas. Linnaeus's great French contemporary, Buffon, held views that derived from the alternative medieval philosophy of nominalism. He held that individuals constitute the only reality. We classify them altogether arbitrarily in groups to which we give names (hence 'nominalism'). By the act of naming species, we create them.

opposed positions, Linnaeus and Buffon were brought closer together as their knowledge of actual diversity increased. Linnaeus never abandoned the concept of essences, but he did abandon his earlier claim that 'we count as many species as were created in the beginning'. In its place he put the genus – oddly, because if any category is real it is the species – and suggested that species might have arisen by hybridisation. He was driven to this position by the sheer impossibility of characterising each species by fixed and unvarying essences. Buffon never accepted essences, but he was driven to recognise that the groups into which organisms fall are not arbitrary and man-made. In other words, he admitted that real divisions exist independently of our judgements.

An interesting confirmation of the reality of species, which has only recently been appreciated, is that primitive peoples usually recognise the same species as do modern taxonomists. The process that ensures the uniformity of the members of a species in any one place is that they interbreed; different species remain distinct because their members do not interbreed. The gradual recognition by the naturalists preceding Linnaeus that species are real entities was an important and necessary discovery. However, it did lend support to an essentialist view of the living world, and for that reason hindered the acceptance of evolutionary ideas. For Darwin, the origin of new species was a central problem. Mayr would say that it was *the* central problem, but I am less sure. I think that for Darwin the most important problem was to provide a natural explanation for the adaptation of organisms to their ways of life. However that may be, Darwin certainly understood that an essentialist view of species was incompatible with evolution.

For both Darwin and Alfred Russel Wallace, a critical step in reaching their evolutionary views was the recognition that populations of a given 'species', if geographically isolated from one another (as, for example, were the animals of the Galapagos Islands), may show varying degrees of difference, from that appropriate to 'varieties' of a single species up to a difference sufficient to justify their being placed in different species. If such isolated populations were once seen as incipient species, that spelled the end of the essentialist view. As Mayr says, 'The fixed, essentialistic species was the fortress to be stormed and destroyed; once this had been accomplished, evolutionary thinking rushed through the breach like a flood through a break in a dike.'

Essentialism, however, did not die with Darwin. When Mendel's laws were rediscovered in 1900, the early Mendelians saw themselves as anti-Darwinian. The mutations they were studying usually had striking effects. Each mutation was seen as the origin, at least potentially, of a new species. The relatively minor and apparently 'continuous' variations in size and shape observed within natural populations were regarded by the Mendelians as irrelevant to evolution. Only mutation could alter the essential characteristics of a species, and it could do so without need for natural selection. In contrast, the Darwinists argued that the mutations of the Mendelians were monsters doomed to early death, and that minor variations were the stuff of evolution.

The reconciliation of these different views required a recognition by naturalists that the genetic variation they saw in natural populations was caused by Mendelian genes, and by geneticists that evolution was a process brought about by selection operating in natural populations, even though the selection was operating on variations that originated by genetic mutation. The story of this reconciliation is one that Mayr is well qualified to tell, since he was one of the main architects of the 'modern synthesis' that resulted. However, even the modern synthesis has not seen the end of essentialist thinking in biology. Its latest manifestation is in the 'punctuationist' theory of the palaeontologists Stephen Jay Gould, Niles Eldredge and Steven Stanley, whose views are reminiscent of those of Hugo de Vries and early Mendelians, but rest on a misunderstanding of the fossil record rather than of mutation.

This, reduced from 800 pages to as many lines, is the picture Mayr paints. How far does he succeed? Magnificently, I think, but perhaps not quite in the task he set himself. In his introduction, he explains that he is attempting a history of how scientists have tackled problems. He remarks on the need to avoid writing a Whig history of science, but that is the kind of history he has written. To be fair, I cannot imagine how a man who has striven all his life to understand nature, and who has fought to persuade others of the correctness of his understanding, could write any other kind of history. Indeed, I am prepared to argue that there are worse kinds of history of science to write than Whig histories – for example, histories of the kind written by Gertrude Himmelfarb, based on a misunderstanding of the science whose history is described. After all, if Victorian England really had been the highest peak of civilisation yet reached, and if it really had held in itself the guarantee of continued progress, Macaulay's method of writing history would have had much to recommend it.

Unfashionable as it may be to say so, we really do have a better grasp of biology today than any generation before us, and if further progress is to be made it will have to start from where we now stand. So the story of how we got here is surely worth telling. Of course, we may not all agree about exactly where we are; indeed, as I shall explain in a moment, I do not fully agree with Mayr. More important, a Whig history will seldom give the reader the pleasure that comes from a sudden and unexpected glimpse into an unfamiliar mind from the past. Mayr's history affords a different pleasure: an understanding of a powerful and creative mind of the present, and of its intellectual roots.

In saying that this is Whig history, I risk being misunderstood. Mayr has not peopled the past with tailor's dummies, dressed up to espouse particular causes. He has read his sources and strives to understand them. Again and again, he told me things about the history of my subject I did not know. Some examples may help. I had always picture Aristotle as an essentialist. After all, where else did Aquinas get his ideas? Mayr has persuaded me that Aristotle explicitly rejected the essentialist approach to animal classification. I have thought, and even told my students, that the debt Darwin owed to Sir Charles Lyell was the concept of uniformitarianism. I am now convinced that Lyell's uniformitarianism was a hindrance rather than a help, and that his real gift to Darwin was that he asked the right questions even if he usually gave the wrong answers.

To give a third example, I have long been puzzled by the fact that, after the publication of the *Origin of Species*, the major advances in biology for the next fifty years took place on the Continent, particularly in Germany, and not in England. It is a strange echo of the way in which the fruits of Newton's *Principia* were reaped in France. I once earned the deep hostility of Julian Huxley by suggesting that the fault lay with T. H. Huxley, who persuaded biologists in Britain and America that, after Darwin, the main task was to elucidate phylogenies by the methods of comparative anatomy, whereas in truth it was to develop a theory of heredity. I still have a soft spot for this view, but Mayr convinced me that a more significant reason was the earlier development of professional biology in German universities.

Mayr, then, is fair to his sources. My complaint, if that is the right word, is that he never reveals to me how anyone could have been so unobservant and illogical as to hold essentialist views. He knows that essentialism is synonymous with sin, and so do I. But surely men like Linnaeus, Georges Cuvier, William Bateson and T. H. Morgan were not born in sin. They must have had some grounds for holding the views that they did. It is striking that essentialism is the automatic, unconsidered philosophy of every physical scientist. There exists the hydrogen atom, and no doubt someone, maybe Schrödinger, wrote down the equations that describe its essence. If essentialism works for them, why doesn't it work for us? Am I quite sure that their way of seeing things can never contribute to biology?

Perhaps I would not be so conscious of these reservations if there were not a small part of Mayr's world view that I do not share. Part of his picture is that the naturalists were usually right, and the mathematicians always wrong or, at best, irrelevant. He has a strong case. Darwin was a naturalist, as was Wallace, and he referred mathematical questions to his wrongheaded cousin Francis Galton. I was fascinated to learn that even August Weismann – for me, the greatest evolutionist after Darwin – was a lover of butterflies. Nevertheless, I think that mathematics is crucial for further progress in evolutionary biology. Mayr is on the whole generous to the mathematical geneticists R. A. Fisher, J. B. S. Haldane and Sewall Wright (he adds the name of S. S. Chetverikov), who showed that Mendelian inheritance and Darwinian selection are compatible. But his treatment of evolutionary biology since 1950 seems to me inadequate, because it misses the central role of mathematics in analysing, for example, the evolution of breeding systems, of social behaviour, and of molecules. Mathematics without natural history is sterile, but natural history without mathematics is muddled.

The issue is best illustrated by Mayr's attitude to what he has called 'bean-bag genetics'. Population geneticists often consider the fate of a single gene at a time – or rather, of alternative forms ('alleles') of a single gene – and ask under what circumstances a given allele will increase in frequency in a population. Mayr objects that selection acts on individuals, not on genes, and that individuals are the product not of one gene, but of complex sets of 'co-adapted' genes. This seems to me to be both true and largely irrelevant. A particular gene will increase in frequency, or not, depending on the effects it has on individual fitness, against the background of all other genes present and the environments experienced. It has to be a 'good mixer'. But the nature of genetic transmission is such that each gene which increases in frequency must do so either by chance or on its own merits, and that is what population geneticists assume.

There are, however, two points of view, which one could almost call the 'English' view, deriving from Fisher, and the

'American' view, deriving from Wright. To oversimplify matters somewhat, Fisher thought that each substitution of one gene for another in evolution occurred because it was beneficial, on its own, and that the role of chance events (other than mutation) was slight. Wright thinks that, often, several gene substitutions would be beneficial if they occurred simultaneously, but that each by itself would be harmful. If so, the only way the change can take place is by chance in a small local population. In effect, the English think that evolution is a hill-climbing process, and the Americans that it also involves jumping across valleys.

As a student of Haldane's, I take an impartial view. However, both views are essentially reductionist, and both were first formulated mathematically. There *is* only bean-bag genetics.

These are relatively trivial disagreements. Yet it is characteristic of Mayr's book that the emotion it arouses in me is a wish to argue with him, not about history, but about his scientific views. Essentially, if I dare use the word, that is what his book is about.

3

Symbolism and Chance

Perhaps the hardest concept in science for most of us to think about is that of randomness. The theme of this essay is that the difficulty arises because of a conflict between thinking in terms of probabilities and another kind of thinking which I shall call symbolic, but which could as well be called analogical.

In his book, *Rethinking Symbolism*, Sperber describes how, when he started anthropological field work among the Dorze, he selected topics for further study. He writes,

> Someone explains to me how to cultivate fields. I listen with a distracted ear. Someone tells me that if the head of the family does not himself sow the first seeds, the harvest will be bad. This I note immediately.

And again,

> When a Dorze friend says to me that pregnancy lasts nine months, I think 'Good, they know that.' When he adds 'but in some clans it lasts eight or ten months', I think 'That's symbolic'. Why? Because it is false.

How do people come to believe things which are in some sense 'false', but which nevertheless convey important truths?

15

The answer seems to be that if you tell a child (or, often enough, an adult) a story which is clearly not true in a literal sense, the child does not react by saying to itself 'that is an obvious lie'. Instead, the child seeks for some symbolic meaning in the story, and does so by seeking a formal analogy between the structure of the story and other structures already familiar to it. This, I think, is the main insight which redeems the structuralist approach to anthropology from total incomprehensibility.

To illustrate this point, let me take an example from Sperber. The Dorze assert that leopards are Christians, and therefore that it is unnecessary to guard one's flocks against them on fast days of the Church. (Of course, in practice, the Dorze do not relax their guard on fast days.) Sperber suggests the following interpretation. The Dorze makes sense of the story by forming the analogy, 'Leopards are to hyenas as the Dorze are to their neighbours.' They are led to the analogy by the following facts: the leopard and the hyena are the two common large carnivores in the area; the former, unlike the hyena, kills all that it eats, and eats only fresh meat; the Dorze differ from their neighbours in that they only eat freshly-killed meat; the Dorze, unlike their neighbours, are Christians. Hence, by saying that leopards are Christians, something is being said about the relation between the Dorze and their neighbours.

I do not know how correct his interpretation is, and it does not much matter. The assumptions behind the interpretation, however, are important, and I think they are correct. They are as follows:

(i) Human beings attempt to make sense of 'stories' which they know to be untrue in a literal sense.

(ii) The main method adopted is to seek for formal analogies between the new story and things already known (which may be things accepted as literally true, or other things accepted as only symbolically true).

(iii) The 'truth' so acquired is moral or aesthetic rather than literal.

(iv) The force of a moral belief is greater if it has been

acquired by an individual through an active attempt to perceive meanings in symbolic inputs, rather than (or as well as) by a set of explicit rules such as the ten commandments.

In case it should be thought that this process is peculiar to societies less technological than our own, remember that many of the stories we read aloud to small children are obviously untrue, even to the child, because they are about talking animals, and even (in a series of stories I read to my own children) about talking steam engines.

What is happening in symbolic thinking is that we are making sense out of nonsense; in Sperber's words, 'Symbolic thought is capable, precisely, of transforming noise into information.' The same thing happens when we interpret a Rorschach ink blot. There is nothing surprising about this. It is an inevitable consequence of the way we perceive reality. It is now generally appreciated that in order to perceive things we form hypotheses about what those things might be.

The recognition of structural similarities between things is a primary method of understanding not only myths but also reality. It is a method which reached its highest level of sophistication in China. The Chinese theory of five elements – wood, earth, metal, fire, water – in some ways seems a more promising basis for science than the four-element theory of the Greeks, because it was a theory of transformations, in which each element in the cycle overcomes the next. However, the five elements became the basis of a complex system of analogies; to each element there corresponded a planet, a colour, a taste, a family relationship, a point of the compass (the centre counted as a fifth), and so on. A still more complex system of analogies was developed in the *Book of Changes*, the *I Ching*. Here, every conceivable object or event is allocated to one of the sixty-four hexagrams.

One can see the *I Ching* as a method of classification (Needham suggests that it is a reflection of the bureaucratic mind), and of ascribing symbolic meanings. However, it is based on and helps to reinforce a view of the world so different from that of modern science that it is hard for someone with my background to grasp it. Some idea of that view can be given by two quotations from Needham's *History of Chinese Science*. He writes,

> It was like the spontaneous yet ordered movement of dancers in a country dance of figures, none of whom are bound by law to do what they do, nor yet pushed by others coming behind, but cooperate in a voluntary harmony of wills.

Also,

> If the moon stood in the mansion of a certain constellation at a certain time, it did so not because anyone had ever ordered it to do so, even metaphorically, nor yet because it was obeying some mathematically expressible regularity depending upon such and such an isolable cause – it did so because it was part of the pattern of the universal organism that it should do so, and for no other reason whatsoever.

Needham sees this organic view of the world as one which scientists will have to acquire as they come to grips with biology. This may be so, although it would require a major shift in outlook; at present, biology seems set on a very different course. However, I did not mention the Chinese world outlook because I wanted to discuss this possibility. Rather, I mention it to illustrate one way in which thinking based on a recognition of analogies can develop. In western science, the road has been different. We recognise the analogy between, say, mechanical and electrical phenomena by recognising that they are described by the same differential equations. Mathematics has become the language of analogy in science; it makes possible the recognition of much more complex structural similarities than the classical structuralist $A : B :: C : D$. Of course, much analogical thinking in science, particularly in the early stages of hypothesis formation, is verbal or pictorial rather than mathematical. The aim, however, is usually to replace such loose analogies by mathematical description; then we can kick away the analogical ladder by which we first reached our hypothesis.

It is time that I passed from a general description of symbolic and analogical thinking to the problem of randomness. As it happens, the *I Ching* provides a bridge.

Although it was not its primary purpose, the *I Ching* has been extensively used, in China and more recently in the West, as a device for foretelling the future. A hexagram was selected by a process comparable to the drawing of lots, and its correlations used for divination. This use of what most scientists would regard as a randomising device in divination is, of course, widespread. The petals of a daisy, the tea leaves in a cup, the livers of birds, the cracks in a scapula thrown in the fire, the fall of dice or cards – all have been pressed into service. The logic is unassailable. If indeed nothing happens by chance, if every event, like part of a holograph, reflects the whole, there is nothing unreasonable about the procedure. It is only necessary to learn the nature of the correlations, and divination becomes a possibility.

It should now be clear that the concepts of randomness and of symbolic thinking are incompatible, at least in their more extreme forms. As a definition of randomness, I suggest that to a scientist a random event is an event into whose causes it is not at present efficient to enquire. This leaves open the question of whether the causes are thought to be in principle unknowable, as in quantum theory, or whether, as in theories of mutation in evolution, the causes of mutation can be and are studied in molecular genetics, but are held to be irrelevant to the evolution of adaptations. It also leaves open the question of how a random sequence of events is to be defined mathematically. For the present purpose it is sufficient that a scientist, by treating some set of events as random, is asserting that there is no structure in these events which is relevant to the phenomena he is attempting to understand. Most of us would assert that the distribution of tea leaves in a cup, or the fall of the milfoil sticks in divining by the *I Ching*, is irrelevant to the future.

If the function of symbolic thinking is to transform noise into information, most scientific theories rest on the assumption that, for any particular purpose, some events (and usually most events) are indeed noise, from which no information can be derived. There is no *a priori* reason for regarding the scientific approach as more reasonable; if anything, the opposite is the case. Further, an obstinate refusal to accept some input as random has sometimes been scientifically fruitful; where would we be today if our predecessors had accepted that the wandering of the planets was random?

Our justification, in so far as we have one, for adopting randomness as a component of scientific theories is that such theories have proved to be fruitful. Darwin is interesting here. In his book, *Darwin on Man*, Gruber shows how, when Darwin first sought for a mechanism of evolution, he saw the problem as one of explaining the nature and causes of mutation (i.e. of the *origin* of new variation). In his final theory of evolution by natural selection, the *existence* of variation has in effect become a postulate, not a problem. Darwin's theory does indeed treat mutations as 'random' in the sense defined above, as events into whose causes it was not efficient to enquire. Darwin himself was well aware that an understanding of evolution did ultimately require an understanding of mutation; this may have been one of the reasons for his long delay in publishing. With hindsight we can see that there was no way in which Darwin at that time could have made much progress in a study of mutation, so his achievement did depend on his treating them as random.

Given the universality of symbolic thinking, and its incompatibility with concepts of randomness and probability, it is worth asking whether the relatively slow development of the mathematical theory of probability and its applications arose from this incompatibility. The basic idea of the 'fundamental probability set' emerged *circa* 1550 in Italy in the work of Cardano and Ferrari. A hundred years later Huygens, basing himself on the work of Fermat and Pascal, published the first methodical treatment of the calculation of probabilities. It was not until the middle of the nineteenth century that scientific theories were formulated in which the concept of probability was central.

I find it hard to account for this slow development either by the intrinsic mathematical difficulties, or by the absence of any external demand for the theory. When the theory did develop, it did so in response to questions about games of chance. There was plenty of gambling for money in the ancient world, and a surprisingly accurate knowledge of the probabilities of various throws. However, this knowledge appears to have been empirical; if so, people must have been prepared to spend a long time throwing dice and astragali to get it. The mathematical calculations required to get the information, given a clear idea of how to combine probabilities, would not have been very difficult. The block seems to have been conceptual. However, my remarks are based not on serious research, but on superficial reading. I want only to suggest that the problem would repay further study.

I cannot end without commenting on one other aspect of the conflict between the symbolic and the random. One cannot spend a lifetime working on evolution theory without becoming aware that most people who do not work in the field, and some who do, have a strong wish to believe that the Darwinian theory is false. This was most recently brought home to me when my friend Stephen Gould, who is as convinced a Darwinist as I am, found himself the occasion of an editorial in the *Guardian* announcing the death of Darwinism, followed by an extensive correspondence on the same theme, merely because he had pointed out some difficulties the theory still faces.

Why should this be so? It happens because people expect evolution theory to carry a symbolic meaning which Darwinism manifestly fails to do. All previous societies have had myths about origins. These myths, like the Dorze belief that leopards are Christians, were not intended to be taken literally. The truth they conveyed was a symbolic one, contributing to an understanding of the purpose of the universe and of man's role in it. Darwinism is also an account of origins, but it is intended to be taken literally and not as a myth. As Bernard Shaw said so eloquently in the preface to *Back to Methuselah*, Darwinism is a rotten myth.

4

Science and the Media

During my lifetime, a new profession has arisen – that of
science journalist. Today, the public learn about science, not
in the main from scientists, but from science writers and
from the producers of radio and television programmes. It
was not so fifty years ago. I was taught no science at school,
but by the time I was eighteen I had given myself an
admirable grounding in science by reading Jeans, Eddington,
Haldane, Huxley, Wells, Einstein and Sherrington. All these
men were writing science for the general public, and all,
except H. G. Wells, were working scientists. For any child
trying to do the same job today, the science would be
filtered through the minds of this new profession. What
kind of a job are they doing?

First, there are two reasons why the emergence of the
new profession has been inevitable. One is that science has
become too important to be left to scientists. The impact of
science on the way we live, and on the ways in which we
may die, has become so all-pervasive that scientists can no
longer expect to be left to their own devices. The Wellsian
dream of a society run by scientists and technocrats was
never a sensible one. It is in my view entirely right that the
planning and the applications of science should be matters

of public and not of private debate. The second reason is television. A working scientist can write a newspaper article, an essay or a book in his spare time, but he cannot make a television programme that way.

The second point to make is that, in this country at least, science journalists face an uphill task. Our culture is still deeply unscientific. We have not yet fully broken free of the old snobbery, according to which work is ungentlemanly, and science is work. Given the intrinsic difficulty of the job, the new profession is not doing badly. The presentation of science on radio and television in this country seems to me enormously superior to its presentation in the United States; I have neither the experience nor the linguistic competence to make any wider comparisons. But there are weaknesses, and it is these that I shall spend time on. In particular, the media show an excessive desire to present confrontation and controversy; they tend to concentrate on the social consequences of science rather than on the science itself; and, because science is no longer presented by scientists, it appears as an impersonal and mysterious edifice, and not as something done by human beings.

Science as confrontation

As an example of the presentation of science as confrontation, I shall take my own subject of evolutionary biology. Something very odd has happened during the past five years or so. The public has been persuaded that Darwinism, as an explanation of evolution, has been exploded. I find repeatedly, when discussing my work with non-biologists, that they are under the impression that Darwin has been refuted. Now there is indeed a ferment of debate among evolutionary biologists, but we are no more likely to abandon natural selection than chemists are to abandon the atomic theory. So what has happened?

Let me start with a curious experience. In the summer of 1980 I was in Vancouver, attending the International Conference on Systematic and Evolutionary Biology, which is held every five years. The only mention of evolution I

found in the local press during the meeting was a full-page
article, reprinted from the *Sunday Times*. This was an
account of some experiments by a young Australian, Ted
Steele, claiming to show Lamarckian inheritance – the
inheritance of an acquired character. If confirmed, this would
not have surprised Darwin (who accepted 'the effects of use
and disuse'), but would certainly have made a dent in modern
'neo-Darwinist' views. Steele's work was to receive an
astonishing exposure in the media. He appeared at least
twice on television, and was repeatedly reported in the press.

The reaction of scientists to Steele was varied. Some
dismissed his results out of hand. More rationally, Medawar
recognised that the work should be repeated, and arranged
for Steele to do so in this country. Since that time, a number
of attempts to repeat his results in other laboratories have
failed. It seems likely that the original claims were mistaken.
Steele cannot be blamed for that. But in the meanwhile, the
public have been left with the impression that neo-Darwinism
has been refuted.

The *Sunday Times* has not, so far as I know, published a
middle-page article announcing that neo-Darwinism has been
vindicated – it would, after all, be rather boring. How have
other newspapers treated evolution? The *Times* has a
miserable record. It marked the centenary of Darwin's death
with an article (by a non-biologist) attacking his views, and
a cartoon of Darwin slipping on a banana skin. It introduced
the trial at Little Rock with an article by Fred Hoyle,
announcing that he has doubts about natural selection. As
it happens, I have doubts about quantum theory, but I am
modest enough to know that they probably reflect my
ignorance, and I certainly would not expect the *Times* to
publish them. The *Guardian* is even odder. Its science
reporting is often excellent. But every now and then it carries
an editorial attacking 'mechanistic biology'. This is so
eccentric that it is best treated as part of the *Guardian*'s
charm, along with the misprints.

Television joined in the anti-Darwinian crusade, with an
Horizon programme, 'Did Darwin get it wrong?' This was
a sad mish-mash of genuine scientific controversy (the
gradual *vs.* 'punctuationist' debate), religious prejudice (the

creationists), a gentleman whose only conceivable reason for inclusion was that he was against evolution, and some molecular biologists who hold some fascinating opinions which they did not have time to explain. (It also carried an interview with Steele, but this was followed by an interview with Brent saying he couldn't get the same results.) There was material for at least three programmes here (the punctuationist debate; creationism; the impact of molecular biology on evolution theory), but to mix them up was merely confusing. It left the impression that the producer had wandered round the US thrusting a microphone in front of anyone who would say anything that sounded anti-Darwinian.

The BBC can plead that they have also produced programmes from an evolutionary standpoint – in particular, the admirable *Life on Earth*. This is true, and we should be grateful. But it is also true that this series, excellent as it was, had little to say about the mechanism of evolution.

It may be that I am paranoid, but I am left with the impression that the press of this country, sometimes supported by television, are giving a false picture of the present state of evolutionary biology. In part this is because of the taste for controversy; it is far more interesting to learn that the scientific establishment is wrong, just as, a hundred years ago, it was more interesting to learn that the religious establishment was wrong. I think it is also in part because people do not *want* to believe that they are the product of evolution by natural selection. They would prefer to believe that God created them with some special role in mind. But that is another story.

The social consequences of science

My second worry is that science is being replaced by the social consequences of science. Now it is quite right that we should be concerned with the moral issues raised by in vitro fertilisation, or the ecological effects of modern farming methods, or the effects of microchips on our jobs. But it is important to remember that such questions are no more the whole of science than the problem of maintaining the church roof is the whole of religion. I have the impression that, for *Horizon* in particular, the social consequences of science, which should rightly be one thread, are in danger of becoming the whole rope. I suspect that other contributors to this discussion may take a different view.

My third point concerns the way in which science is being depersonalised. It is an almost inevitable consequence of the emergence of a new profession of science journalists that scientists rarely talk directly to the public, and therefore no longer appear as people, as Haldane, Jeans and Einstein appeared to me as a boy. This is particularly true of television. Let me offer a caricature of a science programme. At intervals, there will be shots of a junior scientist, usually a girl, wearing a white coat and silently pressing the knobs of some wholly incomprehensible piece of apparatus. There will be an interview with a more senior scientist, usually a man, sitting behind a desk; he sounds excited about something, but the interview is too brief, and he himself too unnerved by the television camera, for it to become clear what he is excited about. Any comprehensible message which does emerge will be carried by an invisible voice, which can be anyone from Richard Briers to Tony Benn, provided they are ignorant of the subject, reading a script written by the effectively anonymous producer.

I have two objections to this; it is boring, and it gives the impression that whatever is being said is some kind of revealed and authoritative truth, rather than something dreamed up by the producer, which may or may not be true. The most entertaining programmes are those presented by a visible human being, with human idiosyncracies; Miller's hands, Bellamy's enthusiasm, Attenborough's willingness to get dirty. When I disagreed with something in *The Body in Question* I knew who I was disagreeing with. If this address does nothing else, I hope it will discourage producers from using anonymous voices.

There is, of course, a snag, which the examples I gave bring out: Miller and Attenborough are communicators rather than researchers, and Bellamy must have found that his work for television has cut into the time he has for research. I'm glad he has done so, but few of us would be willing (or perhaps able) to do likewise. But there are many scientists who have something to say. I was myself fortunate that Peter Jones, then working for *Horizon*, was willing to spend a lot of time helping me to say what I wanted to say about the evolution of behaviour. Unless producers are prepared to take this kind of trouble, the old relationship between scientist and public, which existed for men like Haldane, Eddington and Einstein, and before them for Darwin and Davy, will disappear.

Sins of the 'voice over'

To illustrate the sins of the anonymous 'voice over', and the difficulties one can have communicating on television, I shall give an example from my own experience, of an *Everyman* programme, 'Genesis Fights Back'. This was a film about the creationist movement in the US. It showed a Nazi propaganda film in which the spread of the Jewish people in Europe was compared to a plague of diseased rats. The usual voice over explained that this film illustrated how 'Racists have often used evolutionary ideas to support their political claims', and later concluded 'The only type of social organisation which can evolve, let alone work, is one based on kinship, upon ties of blood and race. In other words, nationalistic selfishness is one inevitable outcome of evolution'.

Even after reading the script, I am quite unable to decide whether whoever wrote it actually thinks that Darwinism necessarily leads to antisemitism and the gas chambers. If you watch the whole film, you may think it does not matter anyway, because, immediately following the bit I have just quoted, there is a filmed interview with me, in which I express quite different views about the relationship between Darwinism and racism.

What then am I objecting to? First, I am *not* objecting to the creationists being given a chance to express their views on television. This was quite proper. Indeed, there was a sequence in which two appalling blonde children sang a pop

song with the words 'I'm no kin to a monkey' which I
wouldn't have missed for worlds. Also, I am not objecting
to someone arguing on television that Darwinism leads to
racism. My first objection is to the argument being made
anonymously, particularly because something said by an
invisible voice carries far more authority than if it is said by
someone one can see. My second objection is that, when I
was interviewed for the programme, I had no idea that the
Nazi film was to be shown (nor, I have since learnt, had
the producer). If I had known, I would have spoken quite
differently.

In fact, it doesn't much matter *what* you say when
interviewed for a television programme, unless you have the
strength of mind to insist on being interviewed live. The
producer usually films about fifteen minutes, and uses one.
So, to offer some advice I have never had the sense to
follow, if you are interviewed, and there is one particular
point you want to make, then make that point, and no
other. Otherwise, you will find that the one thing you really
wanted to say has been left out.

.I do not think that in 'Genesis Fights Back' there was
any intention to misrepresent or mislead. My point is that
the widespread television techniques, of a 'voice over' and
of out-of-context interviews which will be later cut, lead
inevitably to misrepresentation.

Despite these reservations, I think it is important that
scientists should learn to communicate on television, because
I think that at their best they can convey scientific ideas
better than anyone else, because they understand them better,
and care about them more. This is certainly true of the
written word. For me, the best writers of popular science
in recent years have been Peter Medawar, Richard Dawkins
and Stephen Gould. It is not an accident that all three are
working scientists.

Science on radio

I find that I have said nothing of science on radio. This is
because radio seems to me to be largely free of those faults

which I have complained about in the press and on television. We are well served, particularly by Radio 4. This may be in part because it is easier to talk to a microphone than to a television camera, but I think it is also because radio producers have been more willing to look for scientists who will do so.

It will still be objected that most scientists are inarticulate. Some are, and some aren't. The most enjoyable 45 minutes I have ever spent watching television was spent watching Feynman talking to a camera. We cannot all be Feynmans. But if, as scientists, we are going to appear on television or radio, there is a hard lesson we have to learn. We have to forget that our colleagues are listening (they probably aren't anyway), and, forgetting, leave out some of those qualifications we would normally put in to cover ourselves. The reason that James Burke is such a deservedly successful presenter of science on television is that he sounds absolutely certain about everything he says. There is no way I shall ever be as certain of anything as James Burke is of everything, but we have to try.

5

Molecules are not Enough

This book contains a collection of essays about biology, most of which have been published before, in varied and often inaccessible places, together with a new concluding chapter on dialectics. The authors have at least four things in common: they are Harvard professors, they have made distinguished contributions to the theory of ecology and evolution, they are dialectical materialists, and they write with wit and insight. Their thesis is that their philosophy is a valuable aid in the practice and understanding of biology. It is not only that Marxism helps in analysing the history and sociology of science; if you are a working biologist, they are saying, Marxism will help you to plan and to interpret the results of research. Crudely, Marxism is good for you. They argue that their own work has been helped by their philosophy. The claim is not only brave but necessary: I would not take them seriously unless they were willing to make it.

I have known and admired both authors for thirty years. During that time, I have worked on many of the same problems that fascinate them. More relevant, I too have been deeply influenced by Marxism, although in my case the start of my second career, as a biologist, coincided with a growing

disillusion with communist politics, and to a lesser extent
with Marxist philosophy. Inevitably, therefore, this review
is in part autobiographical: it is a debate with a past self.

Any discussion of the value of dialectics in biology must
take in Lysenkoism, and it might as well start there. Lysenko
is the millstone round the neck of the dialectical biologist.
The acceptance of his views, under Party pressure, caused
Russian biology damage from which it has not yet recovered,
even thirty years later. Levins and Lewontin adress these
events, as they must. They do not (unlike some Maoists)
claim that Lysenko was right: 'Far from overthrowing
traditional genetics and creating a new science,' they write,
Lysenkoism 'cut short the pioneering work of Soviet
genetics and set it back a generation. Its own contribution
to contemporary biology was negligible.' They also recognise
that Marxist arguments were used to support Lysenko: 'we
cannot dismiss the obviously pernicious use of philosophy
by Lysenko and his supporters as simply an aberration. . . .
Unless Marxism examines its failures, they will be repeated.'

However, they point out, correctly in my view, that
Lysenkoism is not to be explained merely as an error of
Marxist philosophy, or as a crime committed by a bureaucratic government headed by a paranoid lunatic. The state
of Soviet agriculture, the position of academic genetics at
that time, the class differences between academic geneticists
and agronomists, and the political reactions to collectivisation
and to the German invasion, all contributed. It was, they
argue, a misuse of Marxism to apply it in support of
Lysenko: 'Dialectical materialism is not, and has never been,
a programmatic method for solving particular physical
problems.' Instead, they suggest that the philosophy provides
'a set of warning signs' against theories that are too narrow,
mechanical or abstract.

Now I don't think this will quite do. To see why, one
must first understand why at that time Marxists saw
Mendelian genetics as undialectical. The orthodox view
was that genes influence development, but are themselves
unaltered in the process, and hence that the 'Lamarckian'
process of the inheritance of acquired characters is an impossibility. Hence the gene is a metaphysical and undialectical

entity. Even if a sophisticated Marxist would not reject Mendelism on such apriorist grounds alone, his philosophy would certainly issue 'warning signs' against such a theory. I felt that way myself in 1950, and even went to the length of carrying out an experiment, on temperature acclimatisation in the fruitfly Drosophila, which I hoped – I cannot say expected – would demonstrate Lamarckian inheritance (it didn't). Those of my friends who were both Marxists and biologists tended to be similarly hostile to orthodox genetics.

The paradox about genes was cleared up when their molecular structure was discovered. The current view is that it is information that is transmitted in one direction only: from gene to gene and from gene to protein but not from protein to gene. There is nothing implausible about a device that transmits information unidirectionally. A record-player is such a device: you cannot cut a disc by singing at the speaker. But this solution has emerged from a philosophical approach that has nothing to do with Marxism – and is still rejected by some Marxists. Levins and Lewontin point out that a dialectical picture of the relation between organism and environment can be constructed which is fully compatible with the non-Lamarckian nature of inheritance. This may be so, but it is not the point. It is always possible, after the event, to argue that a correct application of some philosophy would have led to the right solution. Unfortunately, when the issue was still open, most Marxists were not on the side of the angels.

I have spent some time on Lysenkoism because it constituted by far the most important application of Marxism to science. One could argue that the real damage was done, not because a group of scientists held erroneous views, but because those views were imposed by a dogmatic bureaucracy. This is true enough, but raises the question of whether Marxist philosophy predisposed that bureaucracy to think itself justified in imposing a theory on the scientific community. In any case, Lysenko must count as a strike against Marxism: are there positive achievements on the other side? The authors suggest that the contribution of Haldane and Oparin to an understanding of the origin of life is such an achievement. I would agree, but I am not sure that Haldane was significantly influenced by Marxism when he had these ideas – they were published in 1932.

Perhaps a better test is to ask how Levins and Lewontin have been influenced in their own scientific work. A different answer must be given for the two men. Levins was a Marxist before he was a biologist, and all his work shows it. His book *Evolution in a Changing Environment*, although it avoids the usual jargon, is the work of a conscious Marxist. I also think that it was a major contribution to ecology. In it, he faces up to the fact that ecology cannot be a science without theories, and yet any theory of ecology that is simple enough to be comprehensible will be too simple fully to reflect reality. To make matters worse, he recognises that ecological theory cannot be constructed without taking evolution into account, and cannot be applied except in a political and economic context. His attempt to construct models that are at once simple, general and 'robust' (i.e. the conclusions are not too sensitive to slight changes in the model) has influenced all subsequent work in the field. It is perhaps ironic that he made extensive use of mathematical techniques borrowed from capitalist economic theory: I cannot criticise because I have done the same. Since that time, he has worked more on applications of ecological theory. The essays in this book on pesticides, on Latin community health, and on applied biology in the Third World, reflect these interests. They illustrate the power of Marxism in the right hands. I have long thought of Levins as a rare example of a scientist whose work has been strengthened by adherence to a philosophy – Marxism or any other – and this book has confirmed that view.

Lewontin is harder to evaluate. He is one of a handful of contemporaries whose work has altered the way we see evolution. But how far has his success been influenced by his present philosophy? Unlike Levins, his career as a biologist was well established before he became a Marxist. Since that time, he has been an outspoken critic of various applications of biology to man, in particular of sociobiology and of hereditarian views about intelligence. But my impression is that the influence of Marxism on his own research has so far been slight. His most individual attitude to biology has been a wish to make it more like physics. In particular, he has hungered after theories that enable the future to be predicted from the past, as the best theories in physics do. This is an admirable, if somewhat optimistic aim, but it is not peculiarly Marxist.

His own work has been largely devoted to answering the following question: what is the precise nature of the genetic variability of natural populations? I was amused, on a recent visit to New Zealand, to meet a very bright Marxist graduate student who expressed astonishment that, when he visited Lewontin's laboratory, he found everyone grinding up DNA. This reaction, although understandable, is unfair. Marxists are materialists, and there is nothing un-Marxist about studying the material basis of heredity, even by grinding it up. But if it is not un-Marxist, neither is it peculiarly Marxist. What is peculiar about Lewontin is not his use of physical techniques in biology, but the questions he has used those techniques to answer. It seems likely that those questions were in his mind long before he became a Marxist.

There are ways, however, in which his biology has been affected. Perhaps the most interesting is illustrated by his book, *The Genetic Basis of Evolutionary Change*. Its closing words are:

> The fitness of a single locus ripped from its interactive context is about as relevant to real problems of evolutionary genetics as the study of the psychology of individuals isolated from their social context is to an understanding of man's sociopolitical evolution. In both cases context and interaction are not simply second-order effects to be superimposed on a primary monadic analysis. Context and interaction are of the essence.

These are impeccably dialectical words, but what are the interactions he wants us to take into account?

This question could have two anwers. Thus suppose we want to know whether one gene at a locus will replace another in evolution. The new gene will have an effect: in the first instance, it will cause the appearance of an altered protein, or perhaps of the same protein at a different time or place in development. But whether this will increase 'fitness' depends on how its effects interact with everything else going on in the organism, and on how the organism interacts with its environment. This is true, but it is also obvious to almost everyone: even Lewontin's favourite enemy, Richard Dawkins, is well aware of it. In fact, however, the last part of Lewontin's book is devoted to a different kind of interaction. In a chapter entitled 'The Genome as the Unit of Selection' he argues that genes at different loci (i.e. at different places on a chromosome) will become preferentially linked together in cooperating units, and that it is these units, rather than the individual genes, that are relevant in evolution. Now this *might* be true: that is to say, one can construct models of evolving populations in which it is true. However, observations made after the publication of Lewontin's book suggest that in actual populations it is rarely the case (technically, genes in natural populations tend to be in linkage equilibrium).

What this means is that, for most purposes, it is correct to think of evolution occurring 'one gene at a time'. Of course, since new genes only succeed through their interactions with all others, we expect the genes in a given organism to work together to produce a functional whole. But it seems that is is usually *not* the case that groups of cooperating genes are held together, so that they can collectively spread through a species, replacing an earlier group of genes. By analogy, the genome evolves as a language might evolve if it incorporated one new word at a time, and not as a language would evolve if it incorporated a set of words and grammatical structures as a unit. On this matter, I think dialectics may have led Lewontin to espouse a hypothesis which might have been correct, but which probably is not.

I cannot leave this topic without a digression. A consistent proponent of the idea of cooperating groups of genes has been Ernst Mayr, whose earlier writings helped to educate the generation to which Levins, Lewontin and I belong. No one could be less Marxist than Mayr, but there are other sources for dialectical ideas. I remember once asking him whether the geneticist Richard Goldschmidt had been a Marxist, because his writings were permeated (for the worse, in my view) by dialectics. His reply was to remind me that only illiterate Anglo-Saxons have to get their dialectics from Marx and Engels: he and Goldschmidt had been raised on a diet of Hegel. Although chastened, I am unrepentant. It is often fruitful to think of evolution 'one gene at a time'. Lewontin's closing words are absurdly exaggerated. If you doubt this, remember that, from a practical viewpoint, one of the most important evolutionary events of recent years has been the spread of insecticide resistance, and that this can be well understood as the result of selection acting on single genes.

There is much in these essays that I found infuriating, but more that I found illuminating. Let me mention two of the pleasures. One is the first appearance in print of 'Isadore Nabi on the Tendencies of Motion', which has been circulating in samizdat for many years. It is an account of what might have happened if Newton had been replaced by a committee of statisticians. It is the only genuinely funny scientific article known to me, essentially because it is making a serious point. Why is it, by the way, that statisticians bring the best out in their critics? Remember Barnet Woolf's definition of statistics as that branch of mathematics which enables a man to do twenty experiments a year and publish one false result in *Nature*.

A second pleasure is the analogy they draw between the present sweeping advance of molecular genetics and the advance of a medieval army. The latter left behind unreduced castles under siege, because it lacked the weapons for direct assault. Molecular genetics also leaves unsolved problems in its wake. Most notably, it has left unsolved the problem of how eggs turn into animals. Genes may code for proteins, but how does this enable them to control development? It is now clear to everyone, as it was not ten years ago, that the castle of development must be reduced. But how is it to be done? The commonest view, I think, is that the molecular approach will do the job. Sequence enough genes, and the meaning of the message will at last become clear. The authors' opinion, which I share, is that the phenomena of development will have to be studied at their own level, although they would probably agree with me that the laws of development, once understood, will prove to be reducible to molecular biology in the same sense that the laws of heredity have been so reduced. This seems to me one of the central strategic, and therefore philosophical, problems in biological research today: how is development best studied?

A problem that the authors mention obliquely, but do not address directly, is the following. How far have Hegel's dialectical categories been rendered obsolete by advances in mathematics? One interpretation of dialectical materialism would be as follows. Marx and Engels wished to analyse the behaviour of highly complex systems. At that time, mathematics was adequate only for the description of simple dynamical systems. Therefore they were obliged to borrow from Hegel a set of verbal concepts, such as the negation of the negation and the change of quantity into quality. Today developments in mathematics make reliance on such vague verbal concepts less necessary.

This argument can be made more explicit by considering the change of quantity into quality. We now have a mathematical language for describing such changes. Imagine a dynamical system described by a set of differential equations. If we gradually change the parameters in the equations, the behaviour of the system will also change gradually; for example, if the behaviour is to oscillate, then the period and amplitude of the oscillation will change gradually. But ultimately, as we continue to change the parameters, we reach a threshold, or 'bifurcation', at which the behaviour changes dramatically: for example, the system may cease to oscillate, and start to grow exponentially. This, I take it, is a mathematical description of the change from quantity into quality. When one has played with a few systems of this kind, one has a better feel for how things are likely to behave. Other Hegelian categories may be susceptible to a similar analysis, although I am not sure: even in my most convinced Marxist phase, I could never make much sense of the negation of the negation or the interpenetration of opposites. But I note that Levins and Lewontin use the cybernetic concept of feedback when they discuss contradiction.

I would find it easier to draw this essay to a conclusion if I knew what conclusion to draw. It is easy to argue that all scientific research requires some prior philosophical commitment, and that it is therefore better that the commitment be conscious and explicit. The snag, of course, is that too firm a commitment only too easily leads to the espousal of erroneous hypotheses. This is not an objection just to Marxism: Pearson's positivism was just as effective as Lysenko's Marxism in leading to wrong views about genetics. Whether a scientist is conscious or unconscious of his philosophical position, there is no guarantee against backing the wrong horse, although some philosophies make it easier than others to correct one's errors. At the present time, because of the astonishing success of molecular biology during the past thirty years, the overwhelming tendency among biologists is towards reductionism. I therefore welcome the final conclusion reached by Levins and Lewontin:

> We must reject the molecular euphoria that has led many universities to shift biology to the study of the smallest units, dismissing population, organismic, evolutionary and ecological studies as a form of 'stamp collecting' and allowing museum collections to be neglected. But once the legitimacy of these studies is recognised, we also urge the study of the vertical relations among levels, which operate in both directions.

6

Science, Ideology and Myth

Recently, after giving a broadcast on Charles Darwin, I received through the post a pamphlet, 'Why are there "Gays" at all? Why hasn't evolution eliminated "Gayness" millions of years ago?', by Don Smith. The pamphlet points to a genuine problem; the prevalence of sexually ambiguous behaviour in our species is not understood, and is certainly not something that would be predicted from Darwinian theory. Smith's motive for writing the pamphlet was as follows. He believes that the persecution of gays has been strengthened and justified by the existence of a theory of evolution which asserts that gays are of low fitness because they do not reproduce their kind. He also believes that gays can only be protected from future persecution if it can be shown that they have played an essential and creative role in evolution.

I do not find the evolutionary theory he offers in the place of Darwinism particularly persuasive, although it is neither dull nor silly. However, that is not the point I want to make here. I think he would have been better advised to say:

> If people have despised gays because gayness does not contribute to biological fitness, they have been wrong to do so. It would be as

sensible to persecute mathematicians because an ability to solve differential equations does not contribute to fitness. A scientific theory – Darwinism or any other – has nothing to say about the value of a human being.

The point I am making is that Smith is demanding of evolutionary biology that it be a myth; that is, that it be a story with a moral message. He is not alone in this. Elaine Morgan recently wrote an account of the origin of *H. sapiens* intended to dignify the role of women and of the mother–child bond (a relationship about which Don Smith is silent). Earlier, Shaw wrote *Back to Methuselah* avowedly as an evolutionary myth, because he found in Darwinism a justification of selfishness and brutality, and in Lamarckism a theory which justified free will and individual endeavour.

We should not be surprised at Don Smith, Elaine Morgan and Bernard Shaw. In all societies men have constructed myths about the origins of the universe and of man. The function of these myths is to define man's place in nature, and thus to give him a sense of purpose and value. Darwinism is, among other things, an account of man's origins. Is it to be wondered at that it is expected to carry a moral message?

I have always found it hard to know how to react to myths, symbols and traditions. For this reason, I find it easier to start the analysis by looking at myth and symbol in a society very different from my own. Sometimes, distance lends clarity. An account I found illuminating is Sperber's *Understanding Symbolism*. Sperber is concerned with the general problem of how symbols are interpreted, but he illustrates his argument by examples from his study of an African people, the Dorze.

The first striking feature of his account is how he recognises material as symbolic. To quote, 'Someone explains to me how to cultivate fields. I listen with a distracted ear. Someone tells me that if the head of the family does not himself sow the first seeds, the harvest will be bad. This I note immediately.' And again, 'When a Dorze friend says to me that pregnancy lasts nine months, I think "Good, they know that". When he adds "but in some cases it lasts eight or ten months" I think "That's symbolic". Why? Because it is false.' What meaning do people ascribe to statements which, to an outsider, seems manifestly false?

Sperber answers as follows. When one hears something which seems, in a literal sense, to be untrue, one does not say 'That is an obvious lie'; instead, one seeks a symbolic meaning in the story by seeking a formal analogy between the structure of the story and some already familiar structure. Let me illustrate this by taking one of Sperber's examples. The Dorze assert that leopards are Christians, and therefore that it is unnecessary to guard one's flocks against them on fast days of the Church. Sperber suggests the following interpretation. The Dorze makes sense of the story by forming the analogy, 'Leopards are to hyaenas as the Dorze are to their neighbours'. They are led to the analogy by the following facts: the leopard and the hyaena are the two common large carnivores in the area; the former, unlike the hyaena, kills all that it eats, and eats only fresh meat; the Dorze differ from their neighbours in that they only eat freshly killed meat; the Dorze, unlike their neighbours, are Christians. Hence, by saying that leopards are Christians, something is being said about the relation between the Dorze and their neighbours.

This simple example illustrates a number of points. First, symbolic information is interpreted because we refuse to accept that any input is meaningless. Shown an inkblot, we see bats, witches and dragons. This refusal to accept that input is noise lies at the root of divination by tarot cards, tea leaves, the livers and shoulder blades of animals, or the sticks of the *I Ching*. It may also account for the strangely late development of a mathematical theory of probability, or of any scientific theory with a stochastic element. As Sperber says, 'Symbolic thought is capable, precisely, of transforming noise into information.'

Second, the means whereby we make sense of an initially incomprehensible pattern is to recognise structural similarities between the new pattern and one already understood. This device is widely used in science, and has even itself been formalised as 'cybernetics'. It may well be that the major contribution made by mathematics to science has been, not the drawing of conclusions from premises, but in bringing out the structural identity between disparate systems. The major contribution of structural anthropology has been to bring out the isomorphisms between different ideas, or between ideas and social systems. The meanings of symbols depend primarily on the positions they occupy in such isomorphisms.

Third, and in the present context most important, the function of a myth is to give moral and evaluative guidance rather than technical help. However, I find it hard to discover how far people distinguish between these functions. In the case of leopards, Sperber tells us that the Dorze do not in fact refrain from guarding their flocks on fast days, so it seems that they make the distinction. However, it is easier to make the distinction in the case of a story or myth, which is heard or told, than of a ritual which is performed. If, before going into battle, a man sharpens his spear and undergoes ritual purification (or, for that matter, goes to mass and cleans his rifle), he may regard the two procedures as equally efficacious. Indeed, they may well be so, one in preparing the spear and the other himself. If we regard the former as more scientific, it is only because we understand metallurgy better than psychology.

However, it does seem useful to distinguish procedures whose function is to alter the external world, and those intended to alter our own state of consciousness. Similarly, it is useful to distinguish technical instructions from stories intended to persuade us that certain things are right. Indeed, we take some trouble when educating our children to give hints about which category of information is being transmitted. For example, a surprisingly large proportion of the stories read aloud to children are about talking animals, or even talking steam engines. It is as if we wanted to be sure they are not taken as literally true.

The examples of Don Smith and Bernard Shaw show that we do not find it easy to distinguish science and myth. One reaction to this difficulty is to assert that there *is* no difference. Evolution theory has no more claim to objective truth than Genesis. Many scientists would be enraged by such an assertion, but rage is no substitute for argument. In the last century, it was widely held that the scientific method, conceived of as establishing theories by induction from observation, led to certain knowledge. If that were so, then there would indeed be a way to distinguish science from myth, because the truth of a myth certainly cannot be established by induction. However, Darwin and Einstein have robbed us of that confidence in induction – or have liberated us from that prison. By establishing the mutability of species, Darwin showed that there is not a fixed and finite number of kinds of things in the universe, each with a knowable essence; there is no 'Platonic idea' for each species. But induction can only lead to certainty if there is a finite and knowable set of objects, so that one can check that one's theory is true of every kind. If Darwin demonstrated the impossibility of acquiring certainty through induction, Einstein showed that what scientists had been most certain of – classical mechanics – was at worst false and at best a special case of a more general theory. After this twin blow, sure and certain knowledge is something we can expect only at our funerals.

But it is one thing to say that scientific knowledge cannot be certain, and quite another that there is no difference between science and myth. Popper has told us that it was the impact of Einstein, and in particular the wish to distinguish between Einstein's theory and those of Freud, Marx and Adler, that lead him to propose falsifiability as the criterion for distinguishing science from pseudo-science. A scientific theory, he suggested, has the property that observations can be conceived of which, if they were accepted, would show the theory to be false. In contrast, he suggests, no conceivable pattern of human behaviour could falsify Freudian theory.

Popper's views have been attacked, primarily on the grounds that there are no such things as theory-free observations. Every observation is subject to interpretation, conscious and unconscious. Consequently, there can never be certain grounds for rejecting a scientific theory, and hence the distinction between science and pseudo-science disappears.

This criticism seems to me largely to miss the point. If Popper was claiming that scientific knowledge was certain, then the impossibility of certain falsification would indeed be damaging. But he claims no such thing. He insists only on two things. First, that a scientific theory must assert that certain kinds of things cannot happen, so that the theory is falsified *if* certain observations are accepted, and second, that there is a logical asymmetry, so that a theory can be falsified but cannot be proved true by the acceptance of observations. Later philosophers of science, notably Lakatos, have given a more recognisable picture of how scientists actually behave when confronted by a disagreement between theory and observation, but have not in my view weakened the force of Popper's proposed demarcation criterion.

There is, however, a tide of ideas which would deny the distinction. The emotional force behind this tide derives in part from an entirely proper disgust at some of the consequences of technology in the modern world, and in part from an equally proper wish to treat the ideas of other peoples as of equal value with our own. What is common to these two reactions is the conviction, which I share, that scientific theories are not the only kind of ideas we need. It does not follow that scientific ideas are not distinguishable from other ideas. I want now to seek for the intellectual origins of the belief that science and myth are indistinguishable.

One source of this belief lay, surprisingly, in Marxism, a philosophy which has led to very different interpretations of science. One interpretation was pioneered by J. D. Bernal, after whom this lecture is named. For Bernal, the crucial thing about science was that it made socialism possible by providing the techniques needed to satisfy people's wants. Under capitalism, science is frustrated, because the possibilities inherent in it for improving the human condition are not fully realised. But although he saw science as being distorted under capitalism – for example, by being pressed into the service of military research – he does not appear to have thought that science would be prevented, under capitalism, from making progress towards an understanding of nature. In this, his views coincided with those of Marx himself, who largely excluded science from that set of ideas – for example, about religion, philosophy, law – which he saw as reflecting the class interests of those who held them.

Thus Bernal regarded science as the greatest hope for the future, and indeed sometimes wrote as if science was a being with a will of its own, rather than an activity of individual scientists living in society. However, a different thread within Marxism has led to a different end. In 1931, Hessen wrote an article in which he argued that Newton had not only been influenced by the technical problems of his day (e.g. gunnery, navigation), but that the form his theory took reflected contemporary society. Such a view is perhaps more easily understood and more obviously true when applied to Darwin, whose theory did recognise in the living world the phenomena of competition predominant in the society of his day. Indeed, both Darwin and Wallace stated that it was from the economist Malthus that they borrowed their central concept. Although they may have exaggerated the importance of Malthus in the genesis of their ideas, it seems clear enough that the concept of a struggle for existence in nature occurred more readily to them that it would have done had they lived in a more feudal and hierarchical society. If, then, major scientific theories merely project onto nature features of contemporary society, they have more in common with myths than most scientists would readily accept.

Here, it seems to me, a crucial distinction must be made between the psychological sources of a theory and the testing of it. If Darwin's ideas, or Newton's, were accepted because they were socially appealing, then indeed science and myth would be indistinguishable. But I do not think that they were. They were accepted because of their explanatory power and ability to withstand experimental test. Of course, new ideas in science sometimes come from analogies with society, just as, in biology, they come by analogy with physics and engineering. But what matters for the progress of science is not where ideas come from but how they are treated.

Society influences the development of science, both through the problems which seem worth solving and the resources available for their solution. I have little doubt that society also influences scientists, both as individuals and groups, by making some ideas seem worth pursuing and others implausible or unpromising. For example, my own caution about applying to man ideas drawn from the study of animal societies – contrasting with the enthusiasm of men like E. O. Wilson and Richard Alexander – probably owes more to the fact that my political and philosophical views were formed under the shadow of Hitler than to anything internal to biology or sociology.

There is, however, a caricature inherent in the externalist view of science, a caricature I wish emphatically to reject. This is the idea that we can evaluate a scientific theory by reference to the society in which it was born, or to the moral or political conclusions which might be drawn from it. Once accept that view, and science is dead, as genetics died in Russia in 1948.

Today, the belief that there are no objective criteria whereby one can choose between rival theories (and hence, by implication, that one can allow one's prejudices full rein) derives largely, I think, from the work of Thomas Kuhn, although the conclusion is very far from the one which Kuhn himself would wish to draw. Kuhn sees science as divided into 'normal' and 'revolutionary' periods. In a period of normal science, members of a scientific community agree about what assumptions can be made, what problems are worth solving and what will count as a solution, what experimental methods should be used, and, most important, they share a 'paradigm', or set of exemplary solutions to problems which can be used as a standard. Revolutions occur when (usually as a result of long-continued failure to solve certain problems within the accepted frame) a fundamentally new set of assumptions and procedures replaces the old.

All this seems to bear at least some resemblance to reality. (It also bears some resemblance to Popper's remark that 'There is much less accumulation of knowledge in science than there is revolutionary changing of scientific theories'.) The crucial difficulty lies in Kuhn's account of how one paradigm replaces another. He speaks of a 'paradigm debate' in which the proponents 'fail to make complete contact with each other's viewpoints' and in which they 'see the world differently'. Again, there is much in what he says. I have, during my lifetime in science, been engaged in enough arguments in which my opponent and I have been talking right past one another, not to recognise this.

The fallacy – and it is not so much Kuhn's as his interpreters – is to suppose that because two scientists are each unable fully to understand what the other is saying, there is no rational way, given time, of settling the issue between them. It is often the case, at some particular moment, that there is no certain way of choosing between two theories or two methods of approach – science would be a very boring profession if it were not so. However, with the passage of time the choice becomes easier, as one or other approach is more successful in overcoming its difficulties. The trouble is that, in order to work at all, a scientist must often commit himself before the evidence is in. In Darwin's words, one must have 'a theory by which to work'. It is this which gives an air of irrationality to the procedure, and which has led people to suppose that the choice between scientific theories is as arbitrary as that between Arianism and Athanasianism.

How, then, is one to evaluate Kuhn's concept of the incommensurability of scientific theories? It is certainly true that different groups of scientists often fail to understand one another, and even more often fail to listen to one another. It is also true that scientists often embrace a theory wholeheartedly before there are adequate grounds for doing so. But it does not follow that there are no good grounds for choosing between theories, or that the choice ultimately depends on the eloquence of the rival partisans. However committed some people may become to one camp or the other, there is usually a number of scientists well able to understand what is at issue, and to recognise the relevance of experiments supporting one view or the other. The final verdict often combines the new views and the old, but, if so, it is a rational synthesis and not an arbitrary compromise.

Consider an example. The debate between the Mendelians and the biometricians reached an astonishing degree of mutual incomprehension. It also demonstrated the influence of Pearson's previously held philosophical views. As Norton has pointed out, Pearson had actually understood that Mendelian inheritance could account for the phenomena of continuous variability which he had been studying, but still rejected genes because he held it improper to admit the existence of hypothetical entities. Yet, despite all this, no one today would doubt the utility either of Pearson's statistical methods or of Mendelian genetics.

If theories were genuinely incommensurable, and rational choice between them were impossible, progress in science would not be expected. Kuhn himself accepts the reality of scientific progress, but only in the sense that the explanatory power of scientific theories has increased; he doubts whether it is sensible to say that science draws closer to the truth about what is 'really there'. I shall return to this point in a moment.

Before leaving Kuhn, I want to suggest that despite his insights, his insistence on a distinction between normal and revolutionary science, and on the incommensurability of paradigms, has been exaggerated. The major scientific revolution during my working life has been the rise of molecular biology. It has all the features of a new 'disciplinary matrix' in Kuhn's sense – new men, new problems, new experimental methods, new journals, textbooks and culture heroes. But where was the incommensurability? I was myself raised in the old discipline, and have never mastered the experimental methods of the new. Yet my almost immediate reaction to the Watson–Crick paper was that a mystery within classical genetics had been cleared up. I think I was fairly typical. Those trained in classical genetics sometimes had difficulty in learning the new techniques, but there were few conceptual difficulties and no 'paradigm debate'.

Perhaps the birth of molecular biology was not a Kuhnian revolution. As it happens, I suspect that before we make much progress in developmental biology a bigger conceptual revolution may be needed than in the transition from classical to molecular genetics. All the same, if molecular biology could be born without the full panoply of a paradigm debate, where does that leave the concepts of normal and revolutionary science?

The history of genetics also forces us to look again at Kuhn's suggestion that progress in science is progress in explaining, but not progress in knowing what the world is really like. The change from the concept of the gene as a

Mendelian factor to the gene as a piece of chromosome, and thence to the gene as a molecule of DNA, does look to me like progress in knowing what the world is like. But perhaps that is a question I should leave to the philosophers.

I would not have spent so much time discussing scientific theories and myths if the difference between them was obvious. Indeed, they have much in common. Both are constructs of the human mind, and both are intended to have a significance wider than the direct assertions they contain. Popper suggested falsifiability as the criterion distinguishing them, and I think he was right. However, we can often distinguish them also by their function. It is the function of a scientific theory to account for experience – often, it is true, the rather esoteric experience emerging from deliberate experiment. It is the function of a myth to provide a source and justification for values. What should be the relation between them?

Three views are tenable. The first, sometimes expressed as a demand for 'normative science', is that the same mental constructs should serve both as myths and as scientific theories. It is widely held. If I am right, it underlies the criticisms of Darwinism from gays, from the women's movement, from socialists, and so on. It explains the preference expressed by some churchmen for 'big bang' as opposed to 'steady state' theories in cosmology. Although well-intentioned, it seems to me pernicious in its effects. If we insist that scientific theories convey moral messages, the result will be bad morality or bad science, and most probably both. The danger is most apparent in evolutionary biology. Darwinism *is* an account of human origins. In all previous cultures, accounts of origins have been myths, so that it is to be expected that people will treat Darwinism as a myth. If one accepts the Darwinian account, then it is easy to equate 'natural' with 'successful in the struggle for existence'; if one treats the account as a myth, it is equally easy to equate 'natural' with 'right'. The consequence is that people either embrace Darwinism and draw from it the conclusion that gays are unnatural, social services impolitic, and charity wicked, or they are so disgusted by these conclusions that they embrace Lamarckism whether or not the evidence supports it. The first choice is bad morality and the second is bad science. There is no escape from this dilemma, so long as we insist on treating scientific theories as if they were myths. However difficult it may be to convince ourselves and others that 'natural' is *not* equivalent to 'right', the attempts must be made.

The second view is that we should do without myths and confine ourselves to scientific theories. This is the view I held at the age of twenty, but it really won't do. If, as I believe, scientific theories say nothing about what is right, but only about what is possible, we need some other source of values, and that source has to be myth in the broadest sense of the term.

The third view, and I think the only sensible one, is that we need both myths and scientific theories, but that we must be as clear as we can which is which. In essence, this was the view urged by Monod in *Chance and Necessity*. His case was that there is no place in science for teleological or value-laden hypotheses. Yet, to do science, one must first be committed to some values – not least to the value of seeking the truth. Since this value cannot be derived from science, it must be seen as a prior moral commitment, needed before science is possible. So far from values being derived from science, Monod sees science as depending on values. It is odd that he was almost universally attacked by his critics for holding that one can derive values from science, when in fact he argued precisely the opposite. The misunderstanding may have been partly his own fault, since he went on to argue that molecular biology lent some support to socialism, a type of argument of precisely the kind he had earlier ruled out of court. The parts of Monod's thesis which I want here to support are, first, that values do not derive from science, but are necessary for the practice of science, and second, that we should distinguish as clearly as we can between science and myth.

PART 2

ON HUMAN NATURE

The success of biologists in explaining the evolution of social behaviour in animals led, perhaps inevitably, to the idea that the same methods might be equally fruitful if applied to our own species. This programme was launched by the publication of E. O. Wilson's massive book, *Sociobiology*. The debate has been raging ever since. I tend to find myself disagreeing most strongly with whichever side I talked to last. My contribution has been the largely negative one of criticising the more unreasonable claims made by both sides. Apart from a brief history of the debate, *The Birth of Sociobiology*, the essays that follow are all reviews of books. The only one that I take at all seriously is the rather difficult review written jointly with Neil Warren, of the two books by Cavalli-Sforza and Feldman, and by Lumsden and Wilson. Both these books attempt to formulate mathematical models of cultural inheritance. The former makes no exaggerated claims for the models it develops, but Lumsden and Wilson do make startling claims about the joint effects of cultural and genetic change. Now I take mathematics seriously. It is important if, by mathematical reasoning, one can draw non-obvious conclusions. Other reviewers of Lumsden and Wilson had praised the book, or more often condemned it, without, apparently, making any attempt to follow the mathematical argument. I therefore spent several months trying to understand the maths. If you have difficulties in following our review, I can assure you they are nothing compared to the difficulties I had in understanding the book. In the end, we decided that the mathematical models lend little support to the verbal argument, and certainly fail to demonstrate any synergistic effect between cultural and genetic processes. But our review is hard to follow. Unless you have a special interest in Sociobiology, I would recommend that you skip it, and get a bird's eye view of the subject from the other essays.

7

The Birth of Sociobiology

The word 'sociobiology' means different things to different people. For some, it is simply the study of social behaviour in animals, from ants to apes, from the standpoint of modern evolutionary theory. For others, it is an ideological construction calculated to justify racial inequality and the oppression of women. For still others, it is an attempt to reform and revitalise anthropology by an infusion of ideas from biology. The origin of this confusion lies in the publication, ten years ago, of E. O. Wilson's book, *Sociobiology*.

The major part of this book consisted of a massive review of what was then known about animal societies, together with a summary of some of the ideas which had been proposed to explain their evolution. If that had been all, I doubt whether the book would have achieved the fame – or the sales – that it did. Although a valuable summary, it was not, as I shall explain in a moment, the origin of the ideas that have revolutionised our knowledge of social evolution. But it included also a last chapter claiming that biology was about to revolutionise the human sciences, incorporating them at last into the body of science proper. It was this that offended academic anthropologists and sociologists, who resented the threatened takeover, and also the radical left, who saw in the implied 'biological determinism' a renewed attempt to justify racial, economic and sexual inequality by a claim that the status quo is justified because it is natural, and rooted in our genes.

I shall discuss the takeover bid later, but first, what were the new ideas about the evolution of animal societies? The problem is clear; if evolution occurs by the natural selection of properties that ensure individual survival and reproduction, how can we account for cooperative, and even self-sacrificing, behaviour? It is convenient to start with the publication, in 1962, of an influential book by V. C. Wynne-Edwards, of the University of Aberdeen, called *Animal Dispersion in Relation to Social Behaviour*. Wynne-Edwards' thesis was as follows. Animal populations rarely outrun their food supply and starve, because their numbers are usually regulated behaviourally; animals refrain from breeding before their numbers rise too high. To bring this about, special 'epideictic' displays have evolved that signal to individuals the density of the population. These displays form the basis of social evolution. Wynne-Edwards' great merit was that he saw that, if his argument was to hold, natural selection must be acting so as to favour some *populations* at the expense of others, rather than some *individuals* at the expense of others, as Darwin had supposed. After all, it is the population as a whole that benefits if some individuals refrain from breeding: those individuals that do refrain are reducing their own contribution to future generations.

Now such 'group-selection' thinking was prevalent at the time: for example, it was used to explain the evolution of conventional behaviour in animal contests, of sex and recombination, of the non-virulent nature of many parasites, and even the stability of ecosystems. However, the group-selection assumption was usually unstated, and often unconscious: Wynne-Edwards made it explicit. The result was a vigorous debate, in which the role of group selection was central. Most people, including the present writer, concluded that, except in very special circumstances, individual advantage would override group advantage, and that the phenomena described by Wynne-Edwards could satisfactorily be explained by selection acting on individuals. Where did that leave social behaviour? As it happens, the mechanism that is now seen as central was outlined in a paper by W. D. Hamilton that appeared almost simultaneously with Wynne-Edwards' book, but which received much less attention at the time. Hamilton's idea, foreshadowed by J. B. S. Haldane and by R. A. Fisher, has been made widely familiar by Richard Dawkins' *The Selfish Gene*, but is worth restating.

Suppose that gene A causes a 'donor' animal (or, for that matter, a plant or even a virus) to perform an act X, which has the consequence that the animal itself produces C fewer offspring (C = 'cost'), but that some other 'recipient' animal produces B additional offspring (B = 'benefit'). Suppose also that the recipient has a probability r of carrying a gene that is 'identical by descent' to A – if donor and recipient are unrelated, r = 0; if they are identical twins, r = 1; if they are full sibs, r = $\frac{1}{2}$, and so on. Then the consequence of the act is that the number of A genes in the next generation is reduced by C/2 (the division by 2 arises because the donor transmits the A gene to only half its offspring), and is increased by r B/2. Hence the A gene will increase in frequency provided that rB is greater than C.

This derivation of Hamilton's famous inequality is over-simple: if you think you detect a fallacy, you should refer back to the original paper (*Journal of Theoretical Biology*, vol. 7, p. 1), and I hope you will be convinced that you are mistaken. The essential biological point is that cooperation and altruistic interactions are more likely to occur between relatives. It is, therefore, crucial when studying societies, to determine the genetic relationship between their members. Much recent work has been aimed at precisely this question. It is now well established that all complex animal societies are composed of genetic relatives.

There is, however, a second and simpler idea that has to be borne in mind, but which has sometimes been forgotten in the fascination of kinship theory. This is that cooperation may pay both partners. Let me give a simple example. In lions, a pride consists of two or more males holding a group of females. The males rarely fight over access to the females; whichever male first finds and consorts with an oestrous female is not challenged by the others. The males as a group defend the pride against takeover by others groups of males. This cooperation between the members of a group of males was at first explained by kinship, and indeed they often are genetically related. However, it is not at all uncommon for unrelated males to team up, and to hold a group of females. The reason is clear. Two males have a chance of holding a pride for long enough to produce cubs, and three cooperating males have a still better chance; a single male would have no chance at all. A male which fought with other members of its group would not benefit, even if it won, because it needs the cooperation of others to hold the pride against other groups of males.

It is worth pursuing this example a little further. It pays a male lion to cooperate, so long as its partner cooperates: it would not pay if its partner was uncooperative. If we imagine two kinds of males, 'cooperators', C, and 'defectors', D, the payoffs in terms of offspring might be as follows:

$$
\begin{array}{ccc}
 & C & D \\
C & 4,4 & 0,2 \\
D & 2,0 & 1,1
\end{array}
$$

In this 'payoff matrix', the first of each pair of numbers represents the offspring produced by the male adopting the strategy on the left, and the second by the male adopting the strategy above; for example, if both males cooperate, each produces four offspring. The numbers are imaginary, but their relative values are probably correct. Now suppose that most lions in a population cooperate. Each cooperator can expect on average four offspring, whereas a rare defector 'mutant' can expect only two. Hence to cooperate is an 'evolutionarily stable strategy', or ESS, even if the males are unrelated. But to defect is also an ESS, because, if most lions defect, each can expect an average of one offspring, whereas a rare cooperating mutant would get zero. Thus there two possible stable states. Cooperation is stable if once it becomes common, but how does it get started? The obvious answer is that cooperation first occurs between genetic relatives.

I have now mentioned the central ideas used in the analysis of animal societies. First, societies usually consist of relatives. Secondly, both partners in a cooperative interaction may benefit, so that neither would gain by defection. Thirdly, the concept of evolutionary stability can be used to analyse those cases in which the best thing for an individual to do depends on what others are doing. However, the real progress in sociobiology has depended on studies of actual animal societies, and in particular on measurements of the genetic relationship between members, and of the effects of particular actions on survival and reproduction.

A firmer basis for the human sciences?

What of the application of these ideas to humans? This was a major objective of E. O. Wilson, mainly, I think, because he hoped to base ethics on the relatively firm basis of scientific analysis, rather than on the irrationality of religion. As an objective, I think this mistaken: questions of right and wrong cannot be settled solely by reference to what the world is like, which is the province of science. In any case, Wilson's subsequent attempts to combine biology and the human sciences have not fared well. In *Genes, Mind and Culture*, written jointly with C. J. Lumsden, he argues that there will be a synergism between genetic and cultural change, so that, in a species such as our own, genetic evolution will be particularly rapid. This is probably true, for the following reason. Cultural transmission leads to particularly rapid changes in the environment in which we live, and a change in the environment causes new selection pressures and hence genetic evolution. But this is not the effect that Lumsden and Wilson had in mind. What they proposed is that, if behaviour changes for cultural reasons, then genetic changes causing the same behaviour will also proceed particularly rapidly, so that people with different ways of life will acquire a genetic predisposition towards those ways of life. Unfortunately, the mathematical model they proposed to establish this claim has the required effect only if extremely implausible assumptions are incorporated in it, and it seems that no such synergistic effect exists.

However, it does not follow that evolutionary ideas have no relevance for the human sciences. Even if a genetic explanation for the existence of different cultures in different places is rejected, it can still be argued that there are genetically determined characteristics common to all human beings that are relevant to human culture: Noam Chomsky's proposal of an innate capacity for acquiring language, if accepted, would be an example. After all, we are, among other things, animals, and the characteristics that make us uniquely able to sustain a complex culture presumably evolved by natural selection. If we sustain a complex culture, and chimpanzees do not, the reason is that we are genetically different from chimpanzees. Sociobiologists have, therefore, claimed that, if we understood the genetic nature of people, this would provide a firmer base for the human sciences.

This apparently modest suggestion has been met with violent hostility. Some of the hostility has arisen from a natural reluctance on the part of academic anthropologists and sociologists to be taken over by biology: after all, they might have to learn some genetics! But there is a deeper ground for opposition. Fundamentally, it is to phrase the: 'a gene A causes an animal to perform an act X'. Applied to humans, this suggests that a gene might cause a person to be a mathematician, or a homosexual, or to vote for a particular party. Such 'biological determinism' is seen as scientifically unjustified (which, of course, it is), and socially pessimistic, since it implies that the world's present ills are incurable.

I think that the argument is best illustrated by considering a particular example – the avoidance of incest. The anthropologist Claude Lévi-Strauss asserted that the decisive step in the transition from animal to man was the origin of the taboo against incest. In essence, his argument is that, if men are prevented from mating with women from their own family, they are forced to enter into exchange relationships with others, and the resulting inter-family relations are the origin of social organisation. In response, sociobiologists have argued that our ancestors were probably avoiding mating with close relatives long before they could talk.

Behavioural mechanisms preventing such matings have been found in all the apes and monkeys that have been studied, and in many other mammals. In animals that live in groups, either males or, less often, females leave the group before breeding.

Anthropologists have made three points in reply. First, the rules of kinship which determine which marriages are proscribed, and which encouraged, are cultural, and do not correspond precisely to degrees of genetic relationship. Secondly, incestuous intercourse does take place. Thirdly, the avoidance of incest is reinforced by religion and law, which would be unnecessary if it was ruled out genetically. The argument seems to me to be fascinating, but it is one that I am not competent to pursue. I do, however, have some fairly strong views about how it should be pursued. We shall get nowhere if we take up a predetermined position, depending on our professional loyalties, or philosophical and ideological convictions. The value of the sociobiological approach to humans will be decided by whether it enables people to explain things that cannot be otherwise explained, and to ask questions that would not otherwise be asked, but that turn out to have interesting answers. In other words, a new approach in science should be judged by its fruits, and not *a priori*. The history of my own science of genetics has persuaded me that people who decide issues on the basis of philosophical views are likely to get it wrong. Consider Karl Pearson, whose positivist philosophy led him to deny the existence of genes, and T. D. Lysenko, whose Marxism led him to espouse the inheritance of acquired characters.

For me, however, the human applications of sociobiology are peripheral. The most exciting thing to have emerged is what might be called a 'gene-centred' view of evolution. The central idea is that it is genes, and not organisms or populations, that are replicated, and hence, in Dawkins' words 'The fundamental unit of selection, and therefore of self-interest, is not the species, nor the group, nor even, strictly, the individual. It is the gene, the unit of heredity.' The logic leading to this conclusion is clear, and yet it seems to run counter to the obvious fact that it is, by and large, individual organisms which are the target of natural selection, and which in consequence evolve organs that ensure their own survival and reproduction. Hearts, legs, teeth and kidneys really are organs that ensure the survival of individual organisms. If it is our genes, and not organisms, that replicate, why should it be organisms, and not genes, that are adapted for survival? The answer is that, so long as Mendel's laws are obeyed, a gene can increase in frequency only by making the organism in which it finds itself more likely to survive and reproduce.

The truth of this assertion is brought home to us by the discovery of 'selfish DNA' that does not obey Mendel's laws. In particular, there are 'transposable elements' that can multiply within a single organism. Hence they achieve their success by being more likely than other genes in the same individual to be transmitted, and not by enabling that individual to have more descendants. If such behaviour was the rule, rather than an interesting exception, organisms with complex adaptations would not have evolved. This leads us to a final question. How did it come about that most genes, most of the time, play fair, so that a gene's success depends only on the success of the individual that carries it?

8

Models of Cultural and Genetic Change

The topic of human cultural evolution has been something of a scientific no-man's land since the reaction, early in the century, against the canards of Social Darwinism. Most social scientists, and especially British social anthropologists, have turned their backs on it. If pressed, many of them would argue that there is not much of a topic there: there are only social changes and cultural differences which have little 'evolutionary' character and are superimposed on an 'open' human genotype. With the recent revival of scholarly interest in Marx, other social scientists have embraced a Marxian scheme of social evolution.

A few American anthropologists – White, Sahlins, Service, Steward – kept alive an explicit interest in cultural evolution. Their line of enquiry seems to have led to Marvin Harris's 'cultural materialism', which looks to explain human cultural practices by a thoroughgoing biological and ecological functionalism, but is not especially evolutionary. However these anthropologists, like the purer Marxists, tend to consider only the gross features of cultural evolution, such as the move from hunting to agriculture or the development of the state. Cultural evolution is commonly said to be Lamarckian rather than Darwinian, but there has been surprisingly little effort to work out a precise theory of its principles.

Entering the no-man's land as it were from the other side, evolutionary biologists have begun to extend theoretical consideration to the problems of cultural transmission and cultural evolution. Theoretical biologists have the self-confidence which stems from the possession of a successful theory of biological evolution. They are also ready to put their models into precise mathematical form. There may be further advantages in coming from biology to culture. For instance, biologists seem prepared to take seriously as 'culture' absolutely everything which is not genetically transmitted. Social scientists, on the other hand, wear the blinkers of their disciplines, only two of which – sociology and social anthropology – take 'culture' at all seriously, with the large unfriendly buffer zone of psychology in between them and biology: thus they tend to confine themselves to the grosser holistic aspects of society and culture. Biologists may have another advantage inasmuch as they appreciate the hugeness of the problem that Darwin faced and solved. They are therefore more likely than social scientists to feel optimistic about the chances of a comparable intellectual feat in the study of cultural evolution.

The two books on cultural evolution by Cavalli-Sforza and Feldman, and Lumsden and Wilson (referred to here as CF and LW respectively) have some obvious similarities, and some deeper differences. They have in common that they are concerned with cultural transmission; that is, with the fact that the characteristics of individuals are influenced by contact with others, as well as by genetic inheritance. They also have in common that they develop mathematical models of cultural transmission, and, in order to do this, are forced to take a somewhat atomistic view of behaviour. That is, they assume that individuals can be satisfactorily characterised by the presence or absence of particular traits – for example, the practice of birth control, the wearing of long skirts or the manufacture of hand axes. The aims of the two books, however, are very different, and there is a corresponding difference in the nature of the models developed.

Both Cavalli-Sforza and Feldman have made significant contributions to population genetics, and this is apparent in their approach. Conscious of the central role played by population genetics in evolutionary biology, they have asked themselves whether a comparable role might not be played in the social sciences by an analogous theory of cultural transmission. It is important to emphasise at the outset that their models are only analogues of genetic models; they are not themselves genetic. No assumptions are made about the genetic basis of differences between members of a population, and no estimates made of changes in gene frequencies. A second volume is promised in which these genetic aspects are treated. Their models do take into account differences in reproductive fitness, because these are relevant even if transmission is cultural. Suppose, for example, that a woman is more likely to practise birth control if her mother does so, and tells her about it. Then the rate of spread (or reduction) of the trait in the population would be affected by the biological fitness of those who practice it.

CF are not primarily concerned to discuss the empirical data on the nature of cultural transmission, although they do use the results of a survey on Stanford students to illustrate their models. They are concerned hardly at all whether transmission is by imprinting, conditioning, observation, imitation, or by direct teaching. Their position is that people are in fact influenced by parents, teachers and contemporaries, and that it is a matter for future research to study the nature and strength of these influences. In the meanwhile, it is useful to develop models in which these unknowns are represented by parameters. This has the virtues of forcing one to be explicit about assumptions, of indicating the kinds of results which are possible and their dependence on parameter values, and of showing what needs to be measured if models are to be made more realistic. The book has some resemblance to a textbook of population genetics such as that by Crow and Kimura. The models are less extensive, but more novel; Crow and Kimura were standing on the shoulders of giants, whereas CF have had to develop their own models.

CF are explicitly atomistic. They write 'our approach to

cultural evolution will be through the discussion of specific traits, rather than through some overall description of culture.' They consider traits which can exist in one of two, or a small number, of discrete states, and also traits which can vary continuously. A basic component of their models is a 'transmission rule', which specifies the probability that a particular individual will acquire a particular trait, as a function of the traits possessed by other members of the population. Thus the transmission rule combines in a single mathematical expression (for discrete states, a transmission matrix) the effects of (i) other members of the society – parents, teachers, sibs, peers, etc. and (ii) any bias causing people to be readier to accept some traits than others.

The first set of models analyses 'vertical' transmission, i.e. from parent to child. For a two-state trait, the model is specified by the four probabilities, b_0–b_3, that a child will adopt trait H rather than h, according to the four parental states, $h \times h$, $h \times H$, $H \times h$, and $H \times H$. CF find the parameter values for which the population will become fixed for one state, or will have a stable dimorphism, or will oscillate. Correlations between relatives, and the effects of assortative mating, are calculated. Darwinian selection can be introduced by supposing that the trait adopted influences fitness; as explained above, if transmission is vertical, the spread of a trait is influenced by fitness differences, even if all members of the population are genetically identical. However, CF show that a trait that reduces fitness can nevertheless spread to fixation. Finally, a process analogous to genetic drift will operate in finite populations.

CF then turn to models of 'oblique' (individuals influenced by members of the previous generation other than their parents) and 'horizontal' (individuals influenced by other members of their own generation) transmission. They show that oblique transmission makes an internal equilibrium, with h and H both present, less likely. However, it is important to notice a feature of these models which CF do not emphasise. The vertical transmission model is essentially symmetrical, in the sense that, in any particular application, it would be arbitrary which state one called h and which H; asymmetry could of course be introduced through the values

b_0-b_3. In the models of oblique and horizontal transmission this is not so. An h individual can be converted into an H by contact with an H, but not vice versa. Thus H is like the possession of a skill or of a transmittable disease, or knowledge of a rumour; one does not unlearn a skill because one meets someone who does not have it. Finally, the models are extended to cover migration, multiple-state traits, and continuously-varying traits.

Perhaps the best way of giving some feel for these models is to discuss some applications of them to real cases, although we must emphasise that CF's main aim is to provide general-purpose models rather than to solve specific problems. The cases we will choose are the demographic transition, the evolution of language, and the evolution of stone tools.

After a general description of the demographic transition, CF consider how the fall in fertility might be modelled. The first question is whether all women gradually changed their concept of the ideal family size more or less simultaneously, or whether there was a rather sharp distinction between wishing to have a large family (say five children) and wishing to have a small one (say two children), with different women making the transition at different times. CF suggest that the statistical evidence is in better accord with the latter hypothesis. They therefore suggest that an appropriate model is one of combined vertical (mother to daughter) and oblique (or horizontal) transmission of a two-state trait, with a selective disadvantage ($1 - s = 2/5$) associated with trait H (two children) compared to trait h (five children). They show that maternal transmission alone would not be sufficient to overcome the contrary natural selection, and estimate the strength of oblique transmission needed if H is to go almost to fixation (as seems to have been the case). It is not part of their intention to explain why individual women found H preferable to h. They conclude their discussion by pointing out that the transition could also be treated using equations developed to describe the spread of an endemic disease.

As an illustration of their ideas about migration, CF consider the divergence of languages between populations, using data on Micronesian islands. The analogy here is with stepping stone models in population genetics. Different, non-cognate, words with the same meaning are analogous to different alleles at a locus. An island population with two or more such words is analogous to a genetically polymorphic population. Hence it is possible to propose a measure of linguistic similarity between two populations which is analogous to a measure of genetic similarity. In the genetic case, theory leads to the expectation $\phi_{ij} = \phi_0 \exp(-kd_{ij})$, when ϕ_{ij} is the similarity between populations i and j, d_{ij} the geographical distance between them, and ϕ_0 and k are constants depending on population size and mutation and migration rates. Empirically, genetic data fit this expectation rather well. The linguistic data suggest enormous differences in k, and hence in the ratio of 'mutation' to migration (note that 'mutation' need not mean local invention; it could imply introduction from a distant island). It also suggests that there are more shared cognates in distant islands than the exponential relation predicts. This could arise because different words have different rates of change; CF suggest how this might be tested.

Finally, to illustrate models of selection and drift for a continuously varying trait, CF analyse data on early stone tools. The crucial comparison is of the variability of shape at a given place and time with the shift in mean shape at different times. This comparison, together with CF's model of drift, shows that some stabilising force must have been acting – presumably, stabilising selection arising from the efficiency of the tools as tools.

These three illustrative examples (and others that we could have given) have one characteristic in common. A comparison of formal models and data does rule out the simplest assumption (pure maternal transmission in the demographic transition; equal rates of change of different words; pure drift for the shapes of tools), and indicate what kinds of additional assumptions must be made. In this sense, then, the models can be made to do a job of work.

The examples also show the sense in which CF use the term 'cultural transmission'. For a geneticist, all variance which is not genetic is, by definition, 'environmental'. It would therefore seem logical to treat all transmission which is not genetic as 'cultural'. Since infectious disease is not genetically transmitted by humans, it would qualify as cultural by the above criteria. CF do not go quite so far, although they draw on mathematical models of epidemic and endemic diseases. When, however, they use the unique New Guinea disease Kuru (transmitted by cannibalism) as an example of vertical and oblique transmission, they justify its inclusion as an example of *cultural* evolution because a funeral ceremony is involved.

The aims of LW are different in kind. Rather than present a set of formal models, whose practical application depends in the main on parameters whose values are yet to be discovered, LW are concerned to show that we already know enough about cognitive and developmental psychology to formulate an adequate model of cultural and genetic change. The essential feature of their model is that genetic and cultural evolution are linked by epigenisis – i.e. by individual development. They claim that such a model leads to startling predictions. At least some of these predictions had been made by one of the authors before the models were constructed. In his *Sociobiology*, Wilson discussed the 'multiplier effect', according to which 'A small evolutionary change in the behaviour pattern of individuals can be amplified into a major social effort' (*Sociobiology*, p. 11). As we will explain, this assertion is repeated in the present book, with support from a formal model. The present book also claims to show that cultural transmisison can 'increase the rate of genetic assimilation' (although, as we will also explain, the opposite is also asserted). It is not clear to us whether Wilson believed, when he wrote *Sociobiology*, that such an accelerating effect of culture on genetic change existed, although his use of such phrases as 'positive feedback' and 'autocatalytic threshold' suggest that he may have.

However that may be, a central claim of the present book is that there is a 'thousand-year rule', deducible from the model, according to which the genetic basis of human cultural behaviour is likely to change in the order of one thousand years. If so, then it follows, contrary to the view which Wilson appeared to hold in *On Human Nature*, but agreeing with the view he had expressed earlier in an article in *Man and Beast* (1971; eds. J. F. Eisenberg and W.

Dillon), that cultural and behavioural differences between contemporary human groups which have been isolated for a thousand years are likely to be caused in part by genetic differences between those groups.

The crucial difference between CF and LW, therefore, is that the former present a logical structure, the conclusions from which depend on empirical data which the authors consider to be still largely lacking, whereas the latter are making important assertions about the nature of man and society. For this reason, it was clear to us before we tackled the models in LW that we must decide whether the models do in fact justify the assertions made. Further, since the models will inevitably be opaque to many readers, we ask whether the predictions of the models are in any sense counter-intuitive, or whether they are of a kind which can readily be understood without elaborate mathematical calculations. (We should stress here that a model is not rendered worthless if its conclusions can be so understood. Most models have this property; their value is that they make explicit the assumptions that lead to the conclusions.) In particular, we were concerned to discover whether the models do reveal a non-obvious synergistic effect between genetic and cultural change.

Three main models are analysed by LW under the titles 'gene-culture translation', 'the gene-culture adaptive landscape' and 'the co-evolutionary circuit'. The first of these allows for the effects of genetic differences on behaviour, but not for the effects of behaviour on Darwinian fitness. The second, despite its title, ignores cultural transmission, but does allow for the effects of behaviour on fitness. The third, complete, model allows for the effects of genes on behaviour, of cultural transmission on behaviour, and of behaviour on fitness and hence on gene frequencies. These models will be discussed in turn.

In the model of gene-culture translation, individuals can adopt one of two (or, in principle, more than two) culturgens, C_1 and C_2. Which they adopt depends on their genotype, which may bias them towards one or the other behaviour, and on what others are doing. Thus LW introduce the concept of a 'bias curve' which is similar to CF's 'transmission rules'. Compared to CF, LW have little to say about which particular other members of society (e.g. mother, sib, teacher) influence an individual, and allow only for the frequencies of C_1 and C_2 in the whole population. They consider a finite population (of $N = 25$ in most worked examples), and seek the 'ethnographic curve' which is an analogue of Wright's curve of gene frequency in finite populations. That is, it is a distribution of the frequencies, in a steady state, of populations with varying numbers of individuals adopting C_1 and C_2. Different ethnographic curves are obtained, depending on the strength of any initial bias, on the population size, and on the strength and nature (e.g. copying others, or doing the opposite) of cultural influences. The emphasis is on finding the steady state distribution of kinds of populations, rather than on the rate at which a population in one state might change into another.

The main conclusion as an 'amplification law', corresponding to the multiplier effect discussed above, according to which the effect of culture is to amplify relatively small genetic differences. More precisely, what this means is the following. Suppose people copy what others are doing. Then populations will tend to consist either mainly of C_1-users or mainly of C_2-users. The relative frequencies of the two types of population, at a steady state, will be influenced by the innate bias, common to all populations. It is shown that a relatively small change in the innate bias produces a large change in the relative frequencies of the two kinds of populations; this is what is meant by the 'amplification law'.

Lewontin, in a review in *The Sciences*, has argued that the specific nature of gene-culture translation used by LW was chosen precisely because it gave rise to an amplification law. We have therefore considered the following alternative model, which seemed the simplest which can allow both for an initial bias and for 'culture'. Suppose individuals are born successively into a society in which the frequency of trait A is P. Each individual initially chooses A with probability p; if $p \neq \frac{1}{2}$ there is a bias. The individual then interacts in turn with n random members of the population and, on each occasion, if the member differs from himself, he changes with probability k, and if the member is identical to himself, he does not change. It is easy to show, for any finite n and any $k \neq 1$, the equilibrium state of the population is given by $P = p$. That is, there is no amplification. We do not say that this mechanism is more plausible than the one chosen by LW; it is certainly simpler and it is the first one we thought of.

The significance of the amplification law is perhaps best evaluated by the applications of it that LW propose. There are three, to incest avoidance, to the splitting of Yanomamo villages, and to women's fashions. LW's model of incest avoidance is simple in the extreme. They assume that all people have a strong bias against incest, and that the bias is unaffected by culture, and conclude that most societies avoid incest. We agree that there are reasons to think that, given a choice, people avoid sexual relations with those with whom they were raised when young. However, the unanswered questions are why this bias is so strongly reinforced by cultural means, and why the details of which marriages are proscribed vary from culture to culture. The model, by its nature, can say nothing on these questions.

The second application, to Yanomamo village splitting, is much more complex. Leaving and staying are treated as alternative culturgens. It is assumed that individuals have an initial tendency to become leavers or stayers, and also that they copy others. It is also assumed that there is a threshold population size, below which the bias is towards staying and above which it shifts towards leaving. Not surprisingly, it emerges that, for large population size, there is an ethnographic curve with some populations consisting mainly of stayers and others of leavers. This does not seem a helpful model of the process. The conclusions are built into the assumptions, and nothing is said about how long a population, once it becomes large, will wait before it switches to a population of leavers, or about which particular individuals will leave. In fact, as Chagnon showed, groups of relatives leave.

The third example, of women's fashions, is still less illuminating. The phenomenon to be explained is that features of fashion (e.g. skirt length) vary with a period of about one hundred years (the actual graphs suggest irregular fluctuations rather than a regular period, but that is another matter). To explain this, LW assume that there is an innate bias in choice of fashion which changes with a periodicity of one hundred years, independently of any changes in gene frequency or in the choices made by others. Once again it is no surprise that the model exhibits the expected behaviour. LW provide no mechanism to drive the 100-year cycles, so their result lacks any real explanatory power. We might as well explain fatness as being caused by obesity. CF do suggest a plausible mechanism; they note in passing that periodicities could arise if children tended to avoid the fashions adopted by their parents.

To summarise, the gene-culture translation model is less rich than the analogous models in CF, because it says less about the effects of different kinds of transmission, and it is somewhat feeble in application. If it has a point, it is as a preparation for the model of coevolution, described below.

The 'gene-culture adaptive landscape' model is a straightforward population genetics model in which a genotype, instead of determining a particular phenotype, specifies the probabilities that one or other of a range of possible phenotypes will develop. Genetic selection is incorporated, since different phenotypes have different fitnesses. There is no cultural transmission. This model should likewise be seen as a preparation for the full model.

The 'co-evolutionary circuit' model does incorporate both cultural and genetic transmission and natural selection. Ultimately it is on this model that LW must be judged; they claim that it 'produces some remarkable phenomena'. As before, individuals adopt one of two culturgens, C_1 or C_2; for illustration, let us call them not-smoking and smoking respectively. Whether an individual finally adopts C_1 or C_2 depends on an initial bias, determined by his genotype, and on the influence of others; in the cases actually simulated, individuals tend to copy their peers, and there is no parental influence other than genetic. The culturgen finally adopted determines fitness; smoking is bad for you. The population is infinite and random-mating, with separate generations. The model has been simulated to discover rates of genetic change, assuming either no cultural effects, or weak cultural effects. Strong cultural effects are not analysed, because the methods of approximation used in the simulations would not apply if such effects were strong. (In passing, it would not be difficult to simulate a model of this kind with strong cultural effects.)

We now turn to the conclusions, as listed by LW:

(i) 'Pure *tabula rasa* is an unlikely state.' That is, a genotype with equal *a priori* probabilities of smoking or not smoking is not stable against invasion. This seems self-evident. If there exists an allele making its carriers less likely to smoke, and if smoking reduces fitness, then that allele will spread.

(ii) 'Sensitivity to usage patterns increases the rate of genetic assimilation.' That is, an allele making smoking less likely will spread more rapidly if there is a tendency for individuals to copy their peers. This *is* a counter-intuitive result. Thus consider first a pair of alleles A and a, such that AA never smoke, Aa sometimes smoke, and aa always do; that is, there are no cultural effects and extreme genetic determinism. A will replace a, and it will do so as rapidly as possible for the given effect of smoking on fitness; genetic change cannot be more rapid than when heritability is one. This, however, is not the case considered by LW, who suppose that aa individuals have no *a priori* bias for or against smoking ($P = 0.5$; '*tabula rasa*'), whereas AA and Aa have respectively an a priori bias of 0.4 towards smoking and 0.6 to not smoking. Their simulations, comparing the case in which these probabilities are fixed, and unaffected by cultural transmission (so that exactly 40 per cent of AA and Aa individuals smoke), and the case of weak cultural influences, show that, as the generalisation quoted above states, the cultural effects accelerate the change from a to A.

How can this be? It turns out to arise because the model assumes that there is a cost, in fitness terms, of having a nervous system capable of learning. If this cost is omitted, and fitness is made to depend only on whether an individual acquires trait C_1 or C_2, then the introduction of culture slightly slows down genetic change, as expected on common sense grounds. However, if, for the case of cultural transmission only, a fixed cost is levied on all genotypes in the population, this increases the value of the ratio of fitnesses, and so speeds up evolution. This is explained in more detail in an appendix to this review. In our opinion, there is nothing in the model to cause one to change the common sense view that cultural influences of the kind considered here will usually slow down genetic change, rather than accelerate it.

The matter is of some importance, because it is the only non-obvious result to emerge from the model, and on it rests the conclusion that genes and culture interact to accelerate change. To quote LW in this context, 'This catalytic effect might have contributed to the rapid evolutionary increase in human brain size associated with the onset of gene-culture co-evolution.' To which we would reply, 'only if the catalytic effect exists.'

(iii) 'Culture slows the rate of genetic evolution.' This seems to be a direct contradiction of (ii) and to be correct. However, on inspection this generalisation turns out to be based on the following conclusions from the simulations: a '*tabula rasa*', 50–50, genotype is replaced by a 'non-smoking' genotype more slowly than a 'smoking' genotype would be. Obviously so, but what has this to do with culture?

(iv) 'Changes in gene frequency during the co-evolutionary process can nevertheless be rapid.' This is the basis of the 'thousand-year rule' which LW regard as their major finding. In what sense does it follow from the model? Clearly if genotypes influence behavioural phenotypes, if cultural effects are relatively weak (as they are in LW's simulations), and if fitness differences are reasonably large, then the genetic constitution of a population will change substantially in a thousand years. And if not, not. There is no sense in which the thousand year rule follows from the model. It follows from the assumptions of strong selection and high heritability. (In the simulations, individuals adopting the selectively favoured culturgen accumulate 'resources' five times as fast as those adopting the less favoured culturgen.)

(v) 'Gene-culture co-evolution can promote genetic diversity.' This conclusion turns out to depend on the very special fitness functions employed in the simulations, which give rise to frequency-dependent fitnesses. There is no particular novelty in the idea that behavioural phenotypes often have frequency-dependent fitnesses, and that this in turn can maintain polymorphisms; for example, this is the origin of the concept of a mixed ESS.

Our conclusion, then, is that little that is not self-evident emerges from the models, and that the results which LW regard as important, like the 'thousand-year rule', do not depend on the cultural components of the model, but follow directly from the assumption of high heritability and strong selection. However, this rather negative conclusion needs qualifying in two ways. First, the formulation of these models has made it easier to understand what is being assumed, and what is meant by, for example, the 'multiplier effect'. Second, it is not in fact hard to imagine ways in which genetic and cultural change could reinforce one another. Because of oblique and horizontal transmission, cultural change can be very rapid. Culture, however, provides the environment in which genes are selected. Hence a change in culture can cause a (slower) change in gene frequency, and, if the genetic constitution of the individuals composing a society influences the nature of that society (and, on an evolutionary time-scale at least, this is surely true; the difference between human and chimpanzee societies depends on a genetic difference between humans and chimpanzees), then the 'co-evolutionary circuit' is complete. Our complaint is not that genetic and cultural processes have not interacted during human evolution, but that the models in LW do not do much to illuminate the interaction.

So far, we have discussed mainly the mathematical models in LW. However, these models are intended to clarify and support an argument presented in verbal terms. Although in some cases (e.g. the amplification law) the models support views expressed in earlier books by Wilson, the argument is not quite what one might have expected from an acquaintance with his previous work. There is no serious mention of inclusive fitness, kin selection and their possible ramifications. There is no treatment of human sexuality, altruism, attraction, deceit or power, and negligible attention is paid to kinship and socialisation. In fact, human sociobiology to date is found wanting precisely because the channelled ontogeny of mind has been overlooked as the critical link in the coupling of genetic and cultural evolution. This is an attractive notion which promises to implement that old and persistent conviction that some kind of important relation obtains between phylogeny and ontogeny.

Unfortunately the notion of epigenesis as the gene–culture link turns out, in LW, to be rather loose and empty. It is restricted to 'cognitive' development. The substantive examples discussed tend to be sensory and perceptual (colour vision, hearing, taste, smell) rather than fully cognitive. Although LW constantly make theoretical reference to the 'epigenetic rules' which channel development, they present no good examples of a developmental sequence, and state that the epigenetic rules have not been described by developmental psychology. Their few examples tend to be those for which there is a relatively simple link between genotype and behaviour. In none of these cases do they spell out a dynamic of development such that cultural diversity and cultural change can be related to stages, levels or channels in an ontogenetic sequence. Where it is not on thin ice, LW's theory of co-evolution says little more than that some human behaviours which appear to contribute to or constrain cultural traits are under genetic control, and that in the long run cultural evolution must be tracked and limited by genetic evolution. Epigenesis, ontogeny, development play a part in the theory which rests very largely on sheer verbal advocacy.

This is unfortunate when the basic idea, that evolution somehow works through development, is so important and interesting. LW pay no more than passing respect to Waddington's conception of canalisation in an epigenetic field or 'landscape'. They also neglect Piaget's life's work on the ontogeny of cognition, which in many respects could be just what they wanted. Piaget himself was impressed by the implications of Waddington's work for his own, was concerned with the relation between ontogeny and phylogeny, sometimes characterised his theory of cognitive development as one of biological epigenesis, and even tried to show how the cultural development of human thought, as in the history of mathematics, is closely related to the ontogeny of logico-mathematical knowledge. In *On Human Nature*, Wilson found Piaget's tracking of human cognitive growth of great significance (though he misconstrued the meaning of 'genetic' in Piaget's term 'genetic epistemology'). In *Genes, Mind and Culture* LW dismiss Piaget (together with Lévi-Strauss and Chomsky) in one or two sentences, seemingly because his research methods are not experimental: this is strictly true only with a conception of experimental design that would disqualify nearly all the evidence they consider in any case. Piaget's evolutionary theory of cognition has been fruitfully applied to the history of science, to art history, to the evolution of prehistoric tool-making and (in great detail by Hallpike in a book which LW cite inappropriately) to primitive thought. There is a further large body of research in developmental psychology, which LW scarcely touch. Other social scientific evidence is cavalierly sampled, scant and sketchily examined.

Leach, reviewing LW in *Nature* (Vol. 291, p. 267–68), wrote: 'the whole of this elaborate apparatus is devoid of meaning because it assumes that human culture can be broken down into clusters of traits. . .which can be counted and subjected to statistical analysis.' This objection would apply equally to CF. Unhappily, instead of explaining himself, Leach went on to ridicule LW for the occasional turgidity of their style. Ridicule is not an alternative to criticism. It is not *obviously* true that an atomistic analysis of society is doomed to failure. After all, the various parts of an organism – heart, lungs, nervous system and so on – form a much more completely and successfully integrated whole than do the components of a society. Yet a theory of heredity which is essentially atomistic and reductionist has had a lot to say about how organisms evolve. To object to LW or CF on the sole grounds that they are reductionist (and this seems to be the only serious objection raised, for example, by Alper and Lange in an odd article in *Proc. Nat. Acad. Sci.*

USA Vol. 78, pp. 3976–9) will not persuade biologists.

One cannot pretend, however, that atomistic models drawn analogously from genetics do not have huge difficulties to resolve in their application to cultural transmission. In genetic models the 'atoms' are not the traits but the genes which influence those traits; whereas elements of culture are phenotypes with no analogue for the genotype. A possible solution here is to think of cultural traits (Culturgens, Memes) as if they were asexual organisms, with no genotype-phenotype distinction, reproducing themselves and competing for the occupation of men's heads, as Dawkins proposed in *The Selfish Gene*. A problem that looks even harder is the concept of 'cultural fitness', which is at present a hazy and insubstantial analogy with that of 'biological fitness'. As indicated earlier, some cultural traits – for instance, the practice of birth control – have effects on biological fitness. However the rapid increase in the use of the pill did not occur because of differences in biological fitness. It happened because the use of the pill seemed acceptable to many people (and, perhaps, because selling pills is profitable). Until we understand the principle of these selection rules, we do not have a theory of cultural change or evolution.

Together with the problem of cultural fitness there is the matter of cultural complexity. Cultural evolution seems to imply successive moves towards greater 'complexity'; but members of a culture cannot readily be said to be better 'adapted' (other than to their culture) than members of a contemporaneous simpler culture elsewhere. Also relevant here is the interesting possibilty (mentioned by CF, though not original to them) that a Darwinian mode of selection may operate for a 'second order' of cultural objects, whose fitnesses are determined by the nature of individuals, or 'first-order' organisms. However it is also necessary to recognise that culture transcends biological constraints – at least as much as it is shaped by those constraints. Examples are easy: cooking, the wheel, tools, spectacles. This kind of successive transcendance may well be one of the prime movers of cultural evolution, as further improvements replace previous ones. Neither book considers matters this way round. LW consider the processes of human memory at

length while scarcely recognizing that culture prosthetically
makes reliance on individual memory unnecessary in many
respects. They could not have written their book if they had
to rely on their own memories for its material. Culture
makes things easier – or possible at all. And some of its
changes seem more nearly inexorable ('evolutionary') than
others.

In the process the 'environment' itself becomes man-made,
partly a product of human culture and very much so in
modern times, so that 'culture' is something which individuals
have to adapt to as well as a set of adaptations. This hugely
complicates any full theory of cultural evolution, not to
speak of gene–environment relations. The atomistic view
which biologists bring to cultural evolution is of course a
simplification; but we have yet to see if it is the right kind
of simplification. A conception of culture appropriate for
Japanese macaques learning to wash sweet potato might not
easily extend to the human adoption of agriculture or
industry, of Arabic numbers or a phonetic script.

The reservations expressed above begin to approach those
that social scientists will typically have concerning these
books. Mathematical models may help to clarify ideas, but
the correctness of a theory of cultural evolution cannot be
guaranteed merely by formulating it mathematically. There
is no guarantee that theorising by analogy with biological
evolution will be illuminating. However, it is the biologists
who have risked a foray into the no-man's land; and social
scientists, who have nothing much to speak of as a theory
of cultural evolution, ought not to carp and mock. We can
see whether atomism and biological analogy fail only if they
are tried; and if they fail, we shall be the better placed to
see what goes in their stead. A satisfactory theoretical
treatment of human evolution will have to avoid the
disciplinary deformations of both biology and the social
sciences, each of whose abstractions tend to treat the other
as extra-theoretical and ad hoc. The onus falls on social
scientists, therefore, to take CF and LW and the problem
of cultural evolution seriously.

Appendix

Common sense suggests that cultural effects are likely to slow down genetic change. Why then does LW's co-evolution model lead to the opposite conclusion?

Let C_1 and C_2 be the favourable and unfavourable traits. Two cases are simulated.

(i) No cultural effects. Genotype *aa* is '*tabula rasa*'; that is, *aa* individuals have equal probabilities of adopting C_1 and C_2. Genotypes *AA* and *Aa* have probabilities of 0.6 of adopting C_1 and 0.4 of adopting C_2. If we write $P(\bar{A})$ and $P(aa)$ for the frequency of trait C_1 among *AA* (or *Aa*) and *aa* respectively, then $P(\bar{A})/P(aa) = 1.2$.

(ii) Cultural effects. Individuals tend to copy one another. The resulting frequencies of C_1 and C_2 depend on the gene frequency, but if, for example, $p(A) = 0.1$, the frequency of trait C_1 in genotype *AA* and *Aa* is 0.6045 and in *aa* is 0.5047. Hence $P(\bar{A})/P(aa) = 1.198$.

Thus culture has only a small affect on the relative frequencies of the favoured trait, C_1, in the different genotypes, and the effect is of a kind one would expect to slow down genetic change, as common sense suggests it should. Yet LW's simulations show that gene frequency change is faster when there are cultural effects. How can this be?

LW have a complex method of calculating the fitnesses, $F(\bar{A})$ and $F(aa)$, of the genotypes. It is not possible to calculate these values from the book, because there are errors in equations (6.25), (6.44) and in the text. However, Dr Lumsden has been most helpful in explaining how these fitnesses were calculated. It turns out that $F(\bar{A})/F(aa) = 1.094$ if there is culture, and 1.0746 if there is not; hence evolution does go faster when there is culture.

It seems odd that such a miniscule difference in $P(\bar{A})/P(aa)$ should produce a larger difference, in the opposite direction, in the relative fitnesses. The reason is that the learning process itself is supposed to be costly in fitness terms. Thus in case (ii), but not in case (i), a constant cost is subtracted from the fitnesses of both genotypes. Hence in case (i) (no

culture), the fitness ratio $P(\bar{A})/P(aa) \simeq 6.5/6 \simeq 1.08$. In case
(ii) (culture), the ratio becomes $(6.5 - 1)/(6 - 1) \simeq 1.10$,
which is larger. Therefore evolution goes faster if there is
culture.

The actual methods of calculating fitnesses are needlessly
complex, and we have omitted many details. However the
essence of the matter is that if one subtracts the same constant
cost from the numerator and the denominator, the ratio is
increased. This is what the accelerating effect of culture on
genetic change amounts to.

It took much time and work to reach this trivial conclusion
– one which, as far as we can tell, LW themselves did not
appreciate. The moral is that models must be simple if they
are to be illuminating. In the co-evolutionary circuit model,
simplicity has been sacrificed without a corresponding gain
in realism.

It is perhaps churlish to be so critical of the model after
receiving so much help from Dr Lumsden in understanding
it. It is a pleasure to acknowledge how promptly and
thoroughly he has answered our queries.

9

Constraints on Human Behaviour

This, the authors tells us, is 'not a work of science; it is a work *about* science'. Its thesis is that, in the coming period, the social sciences will be transformed by contact with evolutionary biology, in the same way as, in the immediate past, biology has been transformed by contact with chemistry. Human nature, like the nature of ants, hunting dogs or primroses, is the product of evolution by natural selection. It does not vanish with the appearance of culture, but remains as a constraint on the kinds of ways in which people can behave, and the kinds of societies which they can construct. Knowing this, he argues, it is both possible and useful to discover the nature of these constraints.

Like Wilson's earlier *Sociobiology*, the book is bound to arouse controversy; indeed, it is intended to do so. I hope that criticism will be of things Wilson does say, and not of things he does not. As most criticism will come from previously entrenched positions, it may help if I briefly describe my own. On many points I agree with Wilson. I do see recent advances in evolutionary biology, particularly in the field of behaviour, both as important in themselves, and as bound to give rise to renewed speculation about the relevance of biology to sociology. I do not think it sensible to see man as the behaviourists see him, a *tabula rasa* on which culture can write what it will. I do regard our higher faculties, such as our ability to construct mathematical arguments or to have religious experiences, as requiring an explanation in terms of natural selection.

Granting all this, however, I was not persuaded by Wilson's *Sociobiology* that biologists have all that much to contribute, for two main reasons. First, even if there are 'human universals' which act as constraints on our behaviour, their nature will be discovered by psychologists and anthropologists; we biologists can only make encouraging noises from the sidelines. Second, human societies change far too rapidly for the differences between them to be accounted for by genetic differences between their members (Wilson, I think, would accept this). Hence, as differences are what we are primarily interested in, there is little an evolutionary biologist can say.

What, then, of *On Human Nature*? First, and most important, I found it entertaining and stimulating. I still have my doubts about how much progress can be made along these lines, but I would like to see it read both by students and professors of sociology. It would stimulate their imaginations to learn, on the one hand, that human societies represent only a sample of the animal societies that are possible, and on the other, that features of society they may have thought unique to man are found in other animals.

Wilson has defined his attitude to 'genetic determination' more carefully in this book. Interestingly, he recommends Waddington's concept of canalisation as a paradigm for thinking about human nature and nurture. When he speaks of a trait as innate, he does not mean that it will develop in a fixed way regardless of the environment. He means only that a trait will develop in a rather constant manner despite wide changes in the environment, while accepting that a sufficiently drastic change in upbringing (or perhaps a sufficiently early one) might produce a profound change in outcome. By saying that characteristics are innate, he means only that we acquire some characteristics very readily and others only with great difficulty.

How then are we to recognise what is innate? Wilson

reminds us that, classically, geneticists do not ask whether characteristics are genetically determined, but whether differences are. The difference between blue eyes and brown is genetic, and between speaking English and French is not, but it is meaningless to ask whether eyes or languages are genetic or environmental, as both require both genes and an environment. This has seemed to me a serious difficulty with the search for innate human universals. Wilson's answer is that we can look at the difference between all humans on the one hand, and all chimpanzees, say, on the other. Thus, if most human beings avoid committing incest this can be regarded as innate, because there are other animals (not, as it happens, chimpanzees) which regularly do commit incest. The fact that incest does occur in our species does not prove that there is no genetically determined tendency to avoid it, because Wilson is not using the term 'genetically determined' to mean 'certain', but only 'likely in most environments'.

This attitude can be further illustrated by his chapter on aggression. It opens 'Are human beings innately aggressive? This is a favourite question of college seminars and cocktail party conversations, and one that raises emotion in political ideologues of all stripes. The answer to it is yes.' Now the reason this opening raises emotion in political ideologues (including this one) is that it will be taken to mean that human beings are aggressive come what may, that war is inevitable, and that it is therefore a waste of time to work for peace. But it turns out that Wilson does not mean anything of the kind. By saying that we are innately aggressive, he means only that we have shown aggressive behaviour, including warfare, in most, but not all, of the cultural environments in which we have so far found ourselves. He emphasises that the word aggression has been used to describe many disparate patterns of behaviour. He ends the chapter with a discussion of how we might circumvent our tendency to be violent towards one another. Given that these are his views, I think that the opening words of the chapter are unfortunate. They will certainly provoke controversy, but it is likely to be of that singularly useless type which takes place between people who do not understand one another. Rather than speak of traits being innate, it would be more sensible to ask: What are the circumstances, genetic and environmental, in which a particular trait appears, and how can it be modified?

Wilson's attitude can perhaps best be illustrated by his treatment of human sex differences. All known human societies show some economic sex role differentiation, which shows considerable bias, in that some roles (for example, making tools) are performed in most societies by men, and others (for example, making pots) by women, although the bias is never absolute. Both the role differentiation and the bias in it call for an explanation. There seem to be two possibilities. The first is that there is a physical difference between men and women (women bear and feed babies; men are stronger and swifter) and the roles adopted are a direct consequence of this. The second is that there are innate cognitive and temperamental differences between men and women predisposing them to these roles. As technology is reducing the importance of purely physical differences, it would be easier to abolish role differentiation, should we so wish, if the first hypothesis is true.

Wilson's position, in brief, is that there probably are slight cognitive and temperamental differences between the sexes, but that sexual exploitation has been exaggerated as society has become more complex. He makes it quite clear that whether such differences exist or not need not determine our aims in present-day society. We could choose to exaggerate existing differences, or to give equal opportunity (in which case, he believes, we would not achieve statistical equality in role performance), or we could deliberately aim at statistical equality. Thus, he makes it clear (as was not clear, at least to me, in *Sociobiology*) that he does not subscribe to the view that what evolution has produced is morally right.

In general, his attitude seems to me sensible. The physical differences between the sexes are clearly a product of natural selection, and provide a clue to the breeding system of our recent ancestors. As an evolutionary biologist, I would find it odd if there were not also psychological differences; if economic role differentiation is ancient, as it may well be, then selection favouring a psychological difference is also ancient. Wilson offers two types of evidence for such innate differences. One, that girls exposed to masculinising hormone in the womb often behave in ways more typical of boys, is relevant and suggestive. The other, concerning behavioural differences between small children, is less persuasive; even if children are treated alike, they know their own sex, and will imitate adult behaviour.

This book will arouse controversy on many separate issues. Some will criticise it on the grounds that any emphasis on man's biological nature will be used to defend the status quo. I think such criticism mistaken. Women have been monstrously misused; it is natural that many of them are angry and frustrated. But the cause of women will not be helped by refusing to think dispassionately about the nature of sex differences.

There are many other issues. To what extent is a causal, scientific approach – whether sociobiological, Marxist, functional or whatever – appropriate to the analysis of human affairs? If the religious faculty, like the excretory, is a mere product of natural selection, where can we look for the basis of our moral beliefs? To me, the most interesting question is how far evolutionary biology can contribute to the human sciences. As I have explained, I am a doubter. But I have been wrong on this issue before. Ten years ago I regarded incest avoidance as an entirely cultural phenomenon; only a bigot could hold this view today. I hope Wilson's book will be read and debated; I hope, too, that during the debate each side will listen to what others are saying.

10

Biology and the Behaviour of Man

Do we really need another critique of sociobiology? In general, probably not, but perhaps we need this one. Kitcher, like everyone else, approaches the problem with prejudices, but he tries harder and more successfully than most to rise above them. Prejudices are inevitable. It is natural for geneticists and evolutionary biologists to hope that their disciplines will throw new light on the human condition, and equally natural for social scientists to resist the threatened takeover. More important for many of us, previous efforts to apply biology to human affairs have too often ended up as justifications for racial, sexual and class inequalities. Kitcher, who grew up in England, has not forgotten that, in the post-war years, schoolchildren were divided at the age of eleven into sheep and goats, and that this division was justified by the leading experimental psychologists of the day. He and I share this experience – he as a tested child and I as a parent of tested children. It has left us cautious about proposals to use biological theory to plan human institutions.

Kitcher, then, is unsympathetic to the claim that evolutionary biology can guide political judgement, and I suspect he was unsympathetic before he started work on this book.

Unlike some other authors, however, he has undertaken a genuine study. He does understand the ideas he is criticising. He has the biological knowledge to evaluate the evolutionary background to sociobiology, and the mathematical ability to analyse the claims made for it. Above all, he presents sociobiology in its strongest and most coherent form, and avoids the easy option of attacking only its more idiotic manifestations.

He distinguishes sharply between the attempt to understand the evolution of social behaviour in animals, and attempts to understand man. He is sympathetic to the former enterprise. Correctly, he points out that there is no special underlying theory: 'There is no autonomous theory of the evolution of behaviour. There is only the general theory of evolution.' It may be that interactions between relatives, and frequency-dependent fitnesses, were more important in the evolution of the behaviour of birds than in the evolution of their wings, but they are not peculiar to behavioural evolution; kin selection and game theory are just as relevant to plant evolution.

There is, of course, good and bad work in animal sociobiology, and Kitcher gives examples of both. The bad, he points out, has two characteristics: data are quoted as supporting some specific hypothesis, without considering alternatives, and the hypotheses themselves are modified after the fact until data and predictions are brought into line. However, his chapter 'Dr Pangloss's Last Hurrah', which takes issue with the 'adaptationist program', seems to me only partly correct. He presents two genetical reasons for not expecting perfect adaptation. The first is that there are genetic systems, even with constant fitnesses, in which selection will not fix the fittest genotype. The simplest is that of heterozygous advantage: if Aa is fitter than AA or aa, selection cannot produce a population consisting entirely of Aa individuals. This is of course true, but is it interesting? If we want to understand why some species does not have the phenotype predicted by theory, this kind of genetic detail is rather unlikely to be the reason. Suppose, for example, we are interested in the shape of vertebrate wings. Aerodynamic theory shows that the optimal shape is usually elliptical. Pterodactyls, however, never had elliptical wings, but no one would explain this by suggesting that, perhaps, only heterozygotes had elliptical wings. The true explanation has to do with the way in which pterodactyl wings were made. Of course, this would be reflected in the absence of certain kinds of heritable variability, but this is not a useful way of thinking about the problem. For phenotypes of the complexity typically discussed by sociobiologists, it is usually better to think at the level of development and physiology than of genetics.

All the same, there will be cases in which this kind of genetic constraint will be relevant. The second kind of genetic constraint discussed by Kitcher, however, seems to me to be a misunderstanding. He points out that when fitnesses are frequency-dependent, the mean fitness of a population may decrease under selection. Therefore, he says, 'proponents of optimisation analyses who show that a certain design would maximise mean fitness may not automatically assume that selection can produce this design'. Now, as Kitcher understands very well, a major thrust of sociobiology has been to show that selection acting at the level of the individual does not necessarily lead to the evolution of characteristics optimal for the population. One of the main reasons for this is that fitnesses are frequency-dependent. Evolutionary game theory was developed specifically to analyse such cases. It is not sociobiologists who suppose that selection 'maximises [population] mean fitness'. It is ironic that the phrase 'Pangloss's theorem' was first used in the debate about evolution (in print, I think, by myself, but borrowed from a remark of Haldane's), not as a criticism of adaptive explanations, but specifically as a criticism of 'group-selectionist', mean-fitness-maximising arguments.

Thus I think that Kitcher is unfair to sociobiologists when he introduces the argument from frequency-dependence. However, it is a rare slip: in general his account is just. I agree that what we now need in sociobiology is a more cautious analysis of data, and a more careful consideration of alternative hypotheses. This will not come easily, for reasons that are as much sociological as scientific. The critique by Gould and Lewontin has had little impact on practitioners, perhaps because they were seen as hostile to the whole enterprise, and not merely to careless practise of it. Theorists like myself are understandably delighted when some set of observations seems to fit their theories. Fieldworkers, equally understandably, are pleased if their data receive a rational explanation. The time has come, however, for editors and referees to place more emphasis on the quality of the data, and the care with which alternative explanations have been considered, and less on success in fitting the data to some particular theory. But, as Kitcher insists, the enterprise is worthwhile, and the best work is of a high standard.

What of man? Clearly, biology must have something to say. Man is an animal, and has evolved by the same processes as other animals. The debate is between those who, while accepting that man is an animal, argue that he is such a peculiar animal that evolutionary biology can have little to say about his social behaviour, and those who think that the study of human societies, just as of ant societies, must be rooted in biology. The second position Kitcher refers to as 'pop sociobiology'. I think this is a pity, for two reasons. First, it gives an image of superficiality and appeal to popular prejudice which, at least sometimes, is quite unfair: it is hard to imagine anyone less 'pop' than Richard Alexander. Second, it gives the wrong impression of what Kitcher himself is doing: he is scrupulous about putting the best interpretation on sociobiological arguments. But I see his difficulty: we do need a term for the application of sociobiology to human beings, and I have no better one to offer.

Kitcher's basic position is that one cannot dismiss pop sociobiology simply by asserting that it assumes genetic determinism and is therefore false, since plausible sociobiological arguments can be developed which do not assume that genes determine behaviour. There is therefore no escape from considering these arguments in detail, and to see if they stand up and deliver any fruit. Kitcher distinguishes two schools of pop sociobiology, those of E. O. Wilson and of Alexander. Wilson's basic argument he sees in the form of a ladder, as follows:

(i) We can plausibly argue that the members of some population, G, would maximise their fitness by exhibiting behaviour B.

(ii) If we observe that members of G in fact do B, we conclude that B became, and remains, prevalent through natural selection.

(iii) Because selection is effective only if there are genetic differences, we can conclude that there are genetic differences between current members of G and their ancestors, who did not do B.

(iv) Because there are genetic differences, and because the behaviour is adaptive, the behaviour will be difficult to modify by altering the social environment.

This is a shortened version of Kitcher's reconstruction. A major part of his book consists of a step-by-step critique. Clearly, the last step is based upon the shakiest grounds: the fact that our ancestors did B in all previous environments is not proof that they will do B in a wholly new one. The first three steps look more secure. Kitcher's most effective criticism here is not of the logical possibility of taking these steps, but of the ways in which they are in fact taken. For example, consider sexual behaviour. He quotes Wilson as espousing the view that evolution will lead to males that are 'aggressive, hasty, fickle, and undiscriminating', and females that are 'coy'. But theory suggests not one evolutionary optimum, but several, and a number of our primate relatives form longterm pair bonds and show extensive male parental care. Hence there is little justification for Wilson's first step onto the ladder.

A second illegitimate way of getting on to the ladder is to apply to animals words which describe some human behaviour. For example, mallard drakes are said to 'rape' ducks. Now it is true that drakes do force copulations on ducks and, by so doing, probably increase their fitness. What is the harm in calling this rape? If you are interested in ducks, rather little, but if you are interested in people, quite a lot. It implies that human rape occurs because it increases the inclusive fitness of the rapist. The contexts in which rape occurs make this implausible. I agree that there is a danger in applying words such as 'aggression', 'incest', 'homosexuality', and so on to animals and man alike, when the behaviours referred to may be quite different. However I do have reservations. The alternative is often to invent a turgid and incomprehensible vocabulary to describe what animals do. I remember our unsuccessful attempt to introduce the term 'kleptogamy' at an ethological congress, because we feared that the Anglo-Saxon alternative might offend our hosts.

A few years ago, I worked through the equations in Lumsden and Wilson's *Genes, Mind, and Culture* and found them to be badly flawed. Kitcher has done a still more thorough job, and come to essentially the same conclusion: his chapter is entitled 'The Emperor's New Equations'. On three occasions Wilson has found it helpful to find a mathematical collaborator. His first two were Robert MacArthur and George Oster: he was third time unlucky.

The second approach to human sociobiology, taken, for example, by Alexander, Irons, Chagnon and Dickemann, is more direct. Man is treated like any other animal. The question asked is as follows: given the social environment, do people behave so as to maximise their inclusive fitness? The answer, it is claimed, is 'yes'. Unlike Wilson's arguments, which seem to me generally ill-formulated and empty of content, this claim is worth taking seriously, even though it is probably false. Kitcher, attacks it on two fronts. First, he asks what proximate mechanism could possibly bring such behaviour about, since it seems to require an unconscious relationship-calculator and fitness-maximiser influencing our conscious actions. If I were Alexander, I would reply that, if the claim is true, then it is up to psychologists to discover the mechanism.

Kitcher's second line of attack is to ask whether people do in fact maximise their fitness. Here, the test case is Dickemann's account of societies practising female infanticide, and other acts not obviously contributing to fitness. Kitcher gives a careful analysis of this case, and develops a mathematical model of it suggesting that the increases in inclusive fitness that Dickemann proposes would not in fact occur. I am not sure whether he is right, but this is where the action is. This school of sociobiologists do say things about real societies that are testable; I find it hard to believe that they are right, but at least they are not vacuous.

This is an admirable book. Kitcher has the necessary background in biology, mathematics and philosophy. He is aware of his prejudices, and does his best to overcome them. This will not be the last word, but it is the best one yet.

11

Tinkering

Pandas are peculiar bears, which spend much of their days munching bamboo. To do this, they strip off the bamboo leaves by passing the stalks between their flexible thumb and the remaining fingers. But how can a panda have an opposable thumb, when in bears the thumb lies parallel to the fingers, and inseparable from them? In fact, the panda does not have a proper thumb at all: it has five parallel digits just like other bears. The apparent 'thumb' is a modification and extension of a small bone in the wrist. For Stephen Gould, this is a particular and fascinating fact, but it is also an illustration of a general principle. The principle is that evolution proceeds by tinkering with what is already there, and not by following the canons of optimal design. Had the panda been designed by the Great Artificer, He would not have been constrained to make its hand by modifying the hand of a bear and would doubless have come up with a more elegant, if less entertaining solution to the problem of stripping bamboo.

This example comes from the first of this series of essays on natural history, the theory of evolution and the history of biology. It is characteristic of something which, for me, and clearly for Gould, is one of the major attractions of biology: the bringing together of particular fact and general theory. Darwin made it possible for us to see nature simultaneously with the eyes of a child and of a philosopher. The book is full of such examples. What deep lessons would you draw from the turtles that migrate 2000 miles to breed on Ascension Island, or the male mite that mates with all his sisters and dies before he is born?

Stephen Gould is the best writer of popular science now active, and this book the most enjoyable I have read for a long time. It has the two essential features of good science writing: it tells me of facts and ideas that are new to me, and it makes me want to argue with the author. Let me first choose some facts, almost at random. Although I am a professional biologist, I did not know about the panda's thumb. I learnt a lot about the Piltdown forgery, and was delighted to find that my long-felt suspicion that Teilhard de Chardin had something to do with it is not entirely without support. I was strangely encouraged by the story of Randolph Kirkpatrick and his crazy theory of Nummulospheres. I defy anyone to read Gould's account of the evolution of Mickey Mouse without feeling better for it.

I can perhaps best give an impression of Gould's thinking if I describe two essays I mean to argue with him about next time we meet. It is a striking fact that, although Darwin and Wallace arrived independently at the idea of evolution by natural selection, Wallace never followed Darwin in taking the further step of asserting that the human mind was also a product of evolution. Gould has an interesting explanation of this difference. It arose, he suggests, because Wallace had a too simplistic view of selection, according to which every feature of every organism is the product of selection, whereas Darwin was more flexible, and recognised that many characteristics are historical accidents or the unselected corollaries of something that has been selected. Now there are features of the human mind which it is hard to explain as the products of natural selection: few people have had more children because they could solve differential equations or play chess blindfold. Wallace, therefore, was driven to the view that the human mind required some different kind of explanation, whereas Darwin found no difficulty in thinking that a mind which evolved because it could cope with the complexity of life in primitive human societies would show unpredictable and unselected properties.

This is an interesting idea, but it leaves something out of account. Darwin, as soon as he had become convinced that evolution had occurred, and before he had conceived of the theory of natural selection, opened a note book concerned with questions of psychology and metaphysics. The only explanation of this is that he felt at once that his theory must apply to man, and knew that this required that he develop a materialist theory of psychology. I do not know why Darwin so readily made the extension to man (although it was characteristic of him to push ideas to their conclusions), but I do not think it could have had anything to do with his views on selection, which had hardly been formulated at the time.

A second essay I want to argue about is called 'A Quahog is a Quahog'. The question discussed in the essay is whether the names given by non-Western people to animals and birds do or do not correspond to the groups recognised by modern taxonomists. The answer seems to be that, so long as we confine ourselves to the species level, the correspondence is remarkably close. Species, to a modern biologist, are reproductively isolated groups – for example, Great Tits, Blue Tits and Coal Tits in Britain. At levels higher than the species (for example, the class Mammalia, the order Artiodactyla or the family Bovidae) there is virtually no correspondence at all.

I think this conclusion is essentially correct, and is an important proof that truth is not as relative as our cultural relativists would have us think. Species were made by nature, not by man. Why then do I want to argue with Gould? Only because of the last paragraph of his essay, which argues that, since species are real entities with discontinuities between them, it follows that evolution has proceeded in jumps and not gradually. The argument is persuasive, but it overlooks an important point. Primitive people who give names to animals do not travel about much. So long as one keeps in the same place, the distinctness of species is indeed a reality, but if one travels about, it largely disappears. Populations of a 'species' from different places are not identical, and are often so different that it is a matter of taste whether one regards them as belonging to one species or two.

I hope it will be obvious that my wish to argue with Gould is a compliment, not a criticism. Popular science should reflect science as it is practised: this means that it should reflect controversy and uncertainty. Anyone familiar with current debates in evolutionary biology will have noticed that my disagreements about Wallace and about Quahogs reflect a disagreement between Gould and myself about evolutionary theory.

The excellence of *The Panda's Thumb* provokes me into making some more general remarks about popular science. I must make clear at the outset that I do not regard popular writing or broadcasting about science as an inferior or unimportant activity. My own education in science, to the age of almost thirty, depended entirely on reading the popular works of men like Julian Huxley, Wells, Haldane, Jeans, Eddington and Infeld. Had I not been inspired by them, I would not later have become a scientist. More than that, the ideas I got from them were profound, not superficial.

The list of writers I have just given, together with a later generation – for example, Medawar, Dawkins and Gould himself – suggests that a successful writer of popular science must have two qualities. First, all except H. G. Wells were practising scientists, and the exception in a sense proves the rule, because Wells would dearly have liked to be. Second, and more difficult to define, all wrote from a highly personal, even prejudiced point of view.

To take the second point first, the need for a personal viewpoint, if harder to define, is easier to understand. Who wants to eat a boiled egg without salt? I have not yet recovered from the pleasure of discovering J. B. S. Haldane's unique mixture of militant rationalism and paradoxical intelligence. Peter Medawar would be a duller writer if he didn't hate all pompous writing, and all philosophy except Popper's. Gould's idiosyncracies are a passion for the quirks of history, and a conviction that a man's science is part of his humanity, and not infrequently influenced by his political, sexual and racial prejudices. He also holds sadly misguided views about the mechanisms of evolution, and fails to share my prejudice that an ounce of algebra is worth a ton of words. These views, whether or not I share them, are an essential ingredient of his success as a writer.

But why should successful popularisers be scientists? Everyone knows that the prose written by scientists is constipated, full of jargon and altogether lacking in grace. Is it not better that science should be transmitted to the public by professional journalists and broadcasters? It is perhaps inevitable that it should be, because it is unhappily true that many of us do write and talk badly, and few are prepared to take the trouble to improve. However, the results are not always fortunate. The danger is best illustrated by that often excellent TV programme, *Horizon*. As the years go by, *Horizon* deals less and less with science, and more and more with the politics, the social consequences and the technical and medical applications of science. Watching *Horizon*, or reading the *New Scientist*, sometimes drives me to feel that the politics of science is an expense of spirit in a waste of shame, that technology is something Mrs Thatcher wants us to do, that the social responsibility of science is an honourable and necessary bore, that the philosophy of science is an entertainment for those who have passed the philosophopause. What matters is science. I suspect that the reason why, for me at least, the best popularisers have themselves been scientists is that, however interested they may be in politics or history or philosophy, their first love is science itself. Stephen Gould is deeply aware of the social setting in which scientists work, but he does really care about the science they do. Like me, he fell in love with dinosaurs when he was a boy. Often he infuriates me, but I hope he will go right on writing essays like these.

12

Boy or Girl

Why, in the great majority of animals, are there equal numbers of males and females? For John Arbuthnot, writing in 1710, it was evidence of the beneficence of God: 'for by this means it is provided, that the species may never fail, nor perish, since every male may have its female, and of a proportionable age.' But while that might do for man, it will hardly do for those many species in which there is no monogamous pair bond and no parental care, and in which one male can fertilise many females, and yet which have an equal sex ratio.

Darwin, in the 1871 edition of *The Descent of Man*, came close to the solution, but ended by speaking of 'the survival of those varieties which were subjected to the least waste. . .by the production of superfluous individuals of either sex'. In other words, he saw the sex ratio as being an advantage for the variety or species, and not for the individual parent producing offspring of a given sex. By the second edition he had recognised his error, and wrote: 'I formerly thought that when a tendency to produce the two sexes in equal numbers was advantageous to the species, it would follow from natural selection, but I now see that the whole problem is so intricate that it is safer to leave its solution to the future.'

In essence, the correct solution was proposed by R. A. Fisher in 1930, in his book *The Genetical Theory of Natural Selection*. Now Eric Charnov has devoted a whole book to the topic, and to the related questions of why some species consist of hermaphrodites instead of separate sexes, and of how hermaphrodites allocate their time and energy between the two sexual functions. His book is intended for professional biologists: it is too full of algebra and of technical terms to be readily accessible to others. The basic ideas, however, although mathematical, are so simple that they should be general knowledge.

The first point to grasp is the one Darwin was grappling with: the 1:1 sex ratio cannot be explained as conferring a benefit on the species as a whole. Even if sex itself is needed to confer genetic variability and evolutionary potential on the population (a proposition which is in any case much debated, but that is another, more perplexing story), one male in ten would be ample to ensure fertilisation. The higher the proportion of females, the faster the population can grow. Why waste material on males? Some animals, like the greenfly on roses and the water fleas in ponds, abandon males altogether during seasons of increase. The females produce daughters by virgin birth, and produce males only when times get hard.

What is needed is an explanation of the 1:1 ratio in terms of individual advantage to the parent producing the children. Fisher argued as follows. Imagine that mothers could choose the sex of their children: which sex would they choose? We must be clear about the criterion on which choice is to be made. Since, in evolutionary terms, the 'choice' made by a female will, ultimately, be determined by her genes, the choice she makes will be that which maximises the number of genes she passes to future generations. Suppose that there were more women then men (because mothers tend to produce daughters). Then a mother who produced only sons would have, on average, more grandchildren; this follows necessarily from the fact that every child has one father and one mother. Therefore, in a population with an excess of females, genes causing mothers to produce sons will spread; in the same way, if there is an excess of males, genes causing mothers to produce daughters will spread. The only stable state will be one with equal numbers of the two sexes. *QED.*

I have written as if the sex of the child was determined by the mother: an exactly similar conclusion emerges if sex is determined by the father. Fisher took matters a little further. Suppose that children of one sex cost more to produce: for example, in red deer male fawns take more milk than females, and a hind which has a son is more likely to miss breeding the next year. Fisher argued that, in such cases, parents should equalise the effort expended on the two sexes, producing fewer of the more expensive sex.

At this point I must mention a rather serious snag, which arises from the way the sex of a child is determined. In mammals, males produce two kinds of sperm in equal numbers, carrying 'X' and 'Y' chromosomes respectively. Females produce eggs carrying an X chromosome. If the egg is fertilised by a Y-bearing sperm it develops as a male, and if by an X sperm as a female. Thus the immediate cause of the 1:1 ratio is that X and Y sperm are produced in equal numbers, and are equally good at reaching the egg. So far, all efforts in domestic animals to separate the two kinds of sperm have failed. Now Fisher's argument hinges on the idea that genes can influence the sex ratio. But if males produce exactly equal numbers of X and Y sperm, and if these sperm are in other respects indistinguishable, there is no way in which genes in a parent can influence the sex of their children.

I take this snag more seriously than most biologists. There is little evidence that parental genes can affect the sex ratio in species with an X–Y mechanism. I would therefore not expect to find species making a fine adjustment of the sex ratio to variations in the cost of the two sexes, although some apparent examples of such adjustment have been reported. However, I cannot accept this as a *general* explanation of the 1:1 sex ratio. If it were not for the fact that, in most species, most of the time, the selectively favoured ratio *is* 1:1, I am sure that natural selection would have found some other way of determining sex; as we shall see, other ways do exist.

Fortunately, there is one group of animals in which we can forget about the fear that the sex ratio is 1:1 only because there is no way of changing it. This group is the hymenoptera, which includes the bees, ants and wasps, and various parasites such as ichneumon flies. In these insects, the female stores sperm in a receptacle after mating, and decides the sex of each egg as she lays it by either fertilising it, in which case it becomes a female, or by not fertilising it, so that it becomes a male. We know that females can in fact select the sex of each child, because some parasites lay female eggs in large host caterpillars and male eggs in small ones: it pays them to do so because large size increases the reproductive success of females more than it does of males – at least in these insects.

The first major advance in sex ratio theory after Fisher was made by W. D. Hamilton, who realised that Fisher had made a tacit assumption about the breeding system. The nature of the assumption is best seen by considering an exception. Suppose that the offspring of a particular female always mate among themselves, brother with sister. Since one male can mate with many females, a mother will maximise the number of genes she transmits to grandchildren if she produces mainly daughters, and only enough sons to ensure that they are fertilised. Such inbreeding is unusual, but it does occur in parasitic wasps which lay many eggs in one host caterpillar; the young pupate on the caterpillar, and mate with each other when they emerge, without dispersing. As expected, Hamilton found that the sex ratio in such species is strongly biased towards females; there can be as many as ten females to one male.

Things get more complicated when more than one female lays eggs in the same caterpillar. This develops into a competitive game between the two females, in which the best sex ratio for each to choose depends on what the other does. Recently Jack Werren, a student of Charnov's, has shown that a female wasp can detect whether a caterpillar has already been parasitised, and adjusts the sex ratio of its eggs accordingly. Further, it adjusts the sex ratio to a different degree, according to the number of eggs it lays, in just the way predicted by theory.

My own interest in these matters was stimulated because I was trying to apply game theory to the evolution of other traits, from fighting behaviour to plant growth. Some years ago, Eric Charnov visited me in Sussex, and we discussed how game theory could be applied in biology. We agreed that the problem of why some species are hermaphroditic, and in others the sexes are separate, was one which ought to be treated by game theory. I did not, however, pursue the matter for over a year. When I did return to the problem, I found that one can offer a very elegant and simple theory predicting which species should be hermaphroditic, and sent a brief account of it to Charnov. A few days later I was surprised to find in my post a letter from him, since there had not been time for him to receive my letter in Utah and to reply. It turned out to contain exactly the theory I had sent him. I am happy to say that we published it jointly, instead of engaging in a row about priorities.

Our explanation of hermaphroditism is mathematical, but it can be partly explained by the concept of a 'law of diminishing returns'. A plant which produced twice as many seeds would probably not produce twice as many offspring, because the seedlings, falling close to the parent plant, would inevitably compete with one another. Similarly, a plant which produced twice as much pollen would not father twice as many offspring. Therefore, a plant which divided its efforts between producing seeds and pollen would pass its genes on to more offspring than would a specialist male or female. In fact, most plants are hermaphrodites, and many of those which have separate sexes have probably evolved that habit as a method of avoiding self-fertilisation. Animals are less often hermaphrodites – no mammals or birds are so. This may be because there is often a law of increasing returns for male investment. A red deer stag which was 10 per cent larger would probably increase his harem size by more than 10 per cent.

Among vertebrates, several kinds of fish are hermaphroditic, but they are usually 'sequential hermaphrodites': that is, they start out as one sex, and switch to the other. For example, there is a wrasse which lives in groups of five to ten females and one male. If the male is removed, the dominant female changes into a male. Although hermaphroditic fish are usually sequential, one group, the hamlets, are simultaneous hermaphrodites. Eric Fischer studied the behaviour of one species, the black hamlet. The fish do not fertilise their own eggs, although they can produce eggs and sperm on the same day. Some hours before dusk, they court in pairs. After some time, one member of the pair lays a few eggs, which the other fertilises. It is then the turn of the partner to lay eggs to be fertilised. In one evening, each fish will lay a number of batches of eggs, usually taking turns to lay.

What is the reason for this curious process of 'egg-trading'? Fischer has analysed it as a game, in which each partner has something to gain and to lose. Why should not one partner lay all its eggs, wait for them to be fertilised, and then fertilise all the eggs laid by its partner? Surely this would save time and trouble? Fischer's answer is that a fish which laid all its eggs would be too open to exploitation. Its partner could fertilise the eggs, and then leave and pair with another fish. In a game in which the aim is to pass as many genes as possible to future generations, a fish which laid all its eggs at once would risk failure, because it would risk having no eggs to fertilise. This illustrates a point which comes up repeatedly in the analysis of animal courtship; sperm are cheap but eggs are expensive.

Charnov's book considers three related questions. In a bisexual species, what ratio of sons and daughters is it best to produce? In a simultaneous hermaphrodite, how should an individual allocate effort between male and female functions (e.g. between seeds and pollen)? In a sequential hermaphrodite, at what age should an individual switch sex? One important message of his book is that these are all the same question. As his title suggests, they are all questions about the allocation of resources to the male and female roles.

There are still many unsolved problems. Lurking in the background is the biggest question of all: why do organisms reproduce sexually at all? Charnov assumes sex, and asks subsidiary questions. Let me finish with one puzzle which has cropped up recently. In most animals, sex is determined genetically – for example, by the presence of a Y chromosome, or by whether the egg is fertilised. In tortoises, turtles and some other reptiles, this is not so. The sex of a turtle is determined, when it is an egg, by the temperature at which it is incubated. Eggs incubated above about 30°C become females, and below that temperature males. This state of affairs is almost certainly not primitive: both fish and amphibia have genetic sex determination. At first sight, it does not seem a sensible arrangement. In some local populations there is an excess of males, and in others of females. We do not understand why this odd arrangement has evolved. Such questions will continue to be a critical testing ground for theories of evolution.

13

Genes and Memes

The Extended Phenotype is a sequel to *The Selfish Gene*. Although Dawkins has aimed his second book primarily at professional biologists, he writes so clearly that it could be understood by anyone prepared to make a serious effort. *The Selfish Gene* was unusual in that, although written as a popular account, it made an original contribution to biology. Further, the contribution itself was of an unusual kind. Unlike David Lack's classic *Life of the Robin* – also an original contribution in popular form – *The Selfish Gene* reports no new facts. Nor does it contain any new mathematical models – indeed it contains no mathematics at all. What it does offer is a new world view.

Although the book has been widely read and enjoyed, it has also aroused strong hostility. Much of this hostility arises, I believe, from misunderstanding, or rather, from several misunderstandings. Of these, the most fundamental is a failure to understand what the book is about. It is a book about the evolutionary process – it is *not* about morals, or about politics, or about the human sciences. If you are not interested in how evolution came about, and cannot conceive how anyone could be seriously concerned about anything other than human affairs, then do not read it: it will only make you needlessly angry.

Assuming, however, that you are interested in evolution, a good way to understand what Dawkins is up to is to grasp the nature of the debates which were going on between evolutionary biologists during the 1960s and 1970s. These concerned two related topics, 'group selection' and 'kin selection'. The 'group selection' debate was sparked off by Wynne-Edwards' book, *Animal Dispersion in relation to Social Behaviour*. Its thesis is that animals regulate their own numbers behaviourally, rather than being passively regulated by food. Wynne-Edwards further suggested that animals have evolved special displays, usually involving social aggregations (blackcock leks, shearwater rafts, the mass aerial dances of mosquitoes), which inform them of their numbers, so that they can respond by breeding if numbers are low and by refraining if numbers are high. He noted that the entity which would benefit was the whole population, which would not outrun its food supply, and not the individual, which would leave more progeny if it continued to breed regardless of numbers. He therefore suggested that the necessary behavioural adaptations had evolved by 'group selection' – i.e. through the survival of some groups and the extinction of others. Most biologists have doubted whether such a process could actually be effective, and have argued that natural selection typically acts by favouring some individuals rather than others, and not some populations rather than others. However, Wynne-Edwards did raise in a particularly clear way the question of the level at which selection acts – individual, population, species or ecosystem.

At almost the same time, W. D. Hamilton raised another question about how natural selection acts. He pointed out that if a gene were to cause its possessor to sacrifice its life in order to save the lives of several relatives, there might be more copies of the gene present afterwards than if the sacrifice had not been made, because relatives might carry copies of the gene inherited from a common ancestor. The suggestion has obvious relevance for the evolution of social behaviour. To model the process quantitatively, Hamilton introduced the concept of 'inclusive fitness'. To understand this, you must first appreciate that scientists use the word 'fitness', as they do 'force', in a technical sense only loosely related to its colloquial meaning. 'Fitness' is a property of a 'genotype' – that is, of individuals of a particular genetic constitution. Crudely, it is the expected number of offspring produced by individuals of a given genotype in a given environment. Hamilton saw that to use fitness in this sense could lead to wrong predictions about how the frequencies of genes in populations would change – i.e. about evolution. He therefore replaced this classical fitness by 'inclusive fitness', which includes, not only an individual's own offspring, but any additional offspring raised by relatives with the help of that individual, appropriately scaled by the degree of relationship: for example, if I (or more precisely, people with a genotype like mine) help my sister to raise a child she would not otherwise have, that raises my inclusive fitness by one half. It has since become a rule of thumb among students of social behaviour to say that animals will behave so as to maximise their inclusive fitness.

Dawkins, while acknowledging the debt we owe to Hamilton, suggests that he erred in making a last-ditch attempt to retain the concept of fitness, and that he would have been wiser to adopt a full-blooded 'gene's eye' view of evolution. He urges us to recognise the fundamental distinction between 'replicators' – entities whose precise structure is replicated in the process of reproduction – and 'vehicles': entities which are mortal and which are not replicated, but whose properties are influenced by replicators. The main replicators with which we are familiar are nucleic acid molecules – typically DNA molecules – of which genes and chromosomes are composed. Typical vehicles are the bodies of dogs, fruitflies and people. Suppose, then, that we observe a structure such as the eye, which is manifestly adapted for seeing. We might reasonably ask for whose benefit the eye has evolved. The only reasonable answer, Dawkins suggests, is that it has evolved for the benefit of the replicators responsible for its development. It is, he says, foolish to argue about whether some behaviour pattern has evolved for the benefit of the individual or of the group, since both individual and the group are vehicles. Although, like me, he greatly prefers individual advantage to group advantage as an explanation, he would prefer to think only of replicator advantage.

I said earlier that Dawkins was interested in evolution, not in the human sciences. Yet, in *The Selfish Gene*, he introduced, perhaps unwisely, the concept of a 'meme'. A typical meme, as he then conceived it, is a limerick. He would now, I think rightly, prefer to use the word 'meme' only for the physical structure in the brain which represents the limerick. The spoken limerick is then the 'phenotypic expression' of the meme – to a geneticist, the appearance and characteristics of an organism are its 'phenotype' as opposed to its 'genotype', or genetic constitution. A meme can replicate, because if I, knowing a limerick, speak it aloud, the consequence is the appearance in your brain of a corresponding meme. Hence, by analogy with biological evolution, we can expect memes to 'evolve' phenotypic effects favourable to their own replication. However, it may make a crucial difference, as Dawkins acknowledges, that memes can replicate only by generating a phenotypic representation of themselves, whereas genes replicate by a direct template process.

Dawkins' meme concept has been criticised on the grounds that an 'atomic' theory of culture is necessarily wrong. This may well prove to be correct, although I am astonished at the confidence with which it is sometimes asserted. Animal bodies show a far higher degree of coherence and functional interrelationship than do human societies, and yet an essentially atomic theory of genetics has had a lot to say about the evolution of animal bodies. However, that is not the defence which, in *The Extended Phenotype*, Dawkins makes of memes. Instead, he defends himself by saying that he was trying to make a logical point – i.e. that whenever we meet entities capable of accurate replication, we can expect them to generate phenotypes ensuring their own survival. He is not making a takeover bid for the human sciences: he is trying to explain to us the mode of existence of replicators.

The Selfish Gene has already had one intriguing impact on mainstream biology: its influence is acknowledged by the authors of the concept of 'selfish DNA'. To explain this requires a brief digression. The classic picture of Menelian genetics is that each individual receives at conception one complete set of genes from each parent. This is true enough, but it has turned out that, in addition to these typical genes, there is a large amount of DNA in our chromosomes which has no obvious function. Much of it is present, not in two copies per cell, as we would expect for typical genes, but in very large numbers of copies. What is all this 'repetitive' DNA doing? No doubt much of it will turn out to perform some as yet unknown function useful to the organism. The novel suggestion is that much of it may be 'selfish', or 'parasitic'. A cell is full of enzymic machinery for replicating DNA. Dawkins suggests that DNA molecules, which, unlike typical genes, contribute nothing to the life of the organism, might nevertheless live inside cells, just as tapeworms live in intestines. As yet the matter is controversial, but it is very much in the spirit of Dawkins' thinking that some DNA should be parasitic.

I know that Dawkins has been much puzzled by the hostility his first book aroused. In *The Extended Phenotype* he attempts to analyse and disarm this criticism. He ascribes it, in the main, to the fact that he is perceived as a 'genetic determinist'. Is he a genetic determinist, and if so, is there any harm in it?

Dawkins is certainly a determinist as far as behaviour is concerned. That is, he thinks that an animal's (or man's) behaviour is the consequence of its (or his) genetic constitution, upbringing and immediate circumstances. He would not deny that an action may sometimes be a matter of chance. What he would deny is the existence of something called 'free will' as an additional cause of behaviour, over and above those already mentioned. It is only fair to say that I agree with him, at least as a matter of assumption. One could only *prove* that actions could be fully accounted for by genes, upbringing and immediate circumstances by doing the full accounting, and that is too far off to be worth considering. But anyone who attempts a causal explanation of behaviour has to make the assumption that behaviour is caused – it is certainly made by Marxists, who have been Dawkins' most outspoken critics.

I have found that if I make such an assertion to anyone but a professional biologist, I am likely to be met with the response: 'But surely you believe in free will?' To this, the answer is that, of course, I believe in free will. For example, I am writing this review of my own free will: no one made me do it. It is just that I do not see free will as an alternative to genetic and environmental causation. To say that I do something of my own free will says only that my mental disposition at the time was the deciding factor, and not physical constraint.

Dawkins, then, is a determinist, and so is every scientist who studies behaviour, even if they don't know it. But Dawkins is not a *genetic* determinist – unless it be the late Cyril Darlington, no one ever was. J. B. S. Haldane started his lecture course on genetics with the words: 'Genetics is that branch of science which studies the causes of the innate differences between fairly similar organisms.' For our present purpose, the essential word in this definition is 'differences'. If we see two people, one with dark hair and one with blond, we can ask: 'Is the difference genetic or environmental?' By 'genetic', we mean that the difference is caused by a difference between the fertilised eggs from which the two people developed. If one of them has dyed their hair, then the difference is environmental. Thus we can ask of a difference whether it is genetic or environmental, but not of a characteristic. To ask, 'Is hair colour genetic?' is, quite literally, meaningless. To have hair, of any colour, requires that you have both genes and an environment.

To be a genetic determinist, then, would mean that one thought that all differences between the members of a species were genetic. Dawkins certainly does not think this: indeed, as far as I can tell he does not even have a bias in favour of emphasising genetic rather than environmental causes. Why, then, is he seen as a genetic determinist? The reason is quite simple. It is that, *when he is thinking about evolution*, he is only interested in differences that are genetic, and he is quite right. If two animals differ for environmental reasons, the difference may affect their chances of survival, but it will not affect the nature of their children, and hence will have no evolutionary consequences. Of course, if one were concerned, for example, with designing an educational system, environmentally-caused differences would be of profound importance. But Dawkins is not planning schools, he is talking about evolution.

Another reason why Dawkins angers people is that he thinks, and writes, in analogies. This is obvious in his very title, the 'selfish' gene. I have heard a distinguished biologist arguing passionately that, of course, genes are not selfish, because they are not self-aware beings, to which alone the term 'selfish' can properly be applied. I found it impossible to respond to his passion. I suppose that Dawkins referred to genes as selfish because he imagined that no one would take him literally. I do not regard genes as self-aware, but, when thinking about evolutionary problems, I sometimes say to myself: 'Suppose I were a gene, would I cause my carrier to do A or B?' I have every intention of going on doing so.

So far, I have discussed ideas already present in *The Selfish Gene*. What is new in *The Extended Phenotype*? In essence, having argued that we should think about the selection of replicators, and not of vehicles, Dawkins now suggests that we should dissolve the vehicle altogether. Consider, for example, a spider's web. It is not part of the spider, but it is as much part of the phenotype coded for by the spider's genes as is the spider itself. And if webs are seen as phenotypic expressions of spider genes, why should we not see the lakes resulting from the beaver's dams as phenotyic expressions of beaver genes? The expression of genes does not end at the boundary of the body.

Applied to spider's webs, this way of seeing things does not seem so strange. Dawkins goes on, however, to suggest that we should sometimes see the actions of one animal as part of the extended phenotype of the genes of another. He is able to make this plausible by quoting examples of the effects of parasites on their hosts. Larvae of the beetle *Tribolium*, infected by a protozoan parasite *Nosema*, fail to metamorphose, but instead continue to grow, reaching twice the size typical of their species, apparently because the parasite synthesises the appropriate insect hormone. Fresh-water shrimps (*Gammarus*), infected with the immature stages of a parasitic worm, *Polymorphus paradoxus*, swim to the surface of the water instead of keeping to the bottom. If the worms are to develop further, their shrimp host must be eaten by a surface-feeding duck, typically a mallard, and it has been shown that mallards are more likely to swallow infected than uninfected shrimps.

In these and other cases, the behaviour of the host is such as to ensure the survival of the genes in the parasite. A more familiar example is a reed warbler feeding a baby cuckoo. Dawkins adds that colleagues with whom he has discussed the extended phenotype repeatedly came up with the same speculations. Do cold viruses cause us to sneeze so as to increase their chance of reaching another host? Does any venereal disease increase libido? These are speculations, but they are natural ones if we accept Dawkins' view.

To me, however, the most original chapter in the book is the last, on 'rediscovering the organism'. For most people, organisms are given, and the problem is to explain why they have genes. Once accept a gene's-eye view of evolution, however, and the question becomes: why do genes character-istically band together in organisms? I cannot here summarise Dawkins' answer: indeed, he does not claim to provide a final answer. I can, however, give some idea of his approach by mentioning some subsidiary questions he discusses. What are the limits of an organism? Obviously, a dog or a man or a pine tree is an organism. But is a stand of nettles, all derived by underground roots from one original seed, one organism or many? If we are going to count all the plants arising from a single seed as one organism, what of a clone of aphids, all descended by virgin birth from a single female, and all genetically identical? Does it matter that in the case of the aphids (or, to give a second example, in a clone of dandelions) each new individual is derived from a single cell, whereas in the stand of nettles this is not so? Does it matter whether the single cell is produced sexually or asexually?

In a sense, these questions are purely semantic: we can use the word 'organism' to mean what we like. However, since we think in words, our choices are important. What Dawkins is really asking is: what is the significance for evolution of these various patterns of growth and reproduction? His answer is that the crucial distinction is whether the new individual arises from a single cell, or from many. Essentially, this is because only in the former case is it possible for genes with interesting effects on development to be favoured by natural selection.

I have left till last what is to me the strangest feature of both books, because I suspect it will not seem strange to many others. It is that neither book contains a single line of mathematics, and yet I have no difficulty in following them, and as far as I can detect they contain no logical errors. Further, Dawkins has not first worked out his ideas mathematically and then converted them into prose: he apparently thinks in prose, although it may be significant that, while writing *The Selfish Gene*, he was recovering from a severe addiction to computer programming, an activity which obliges one to think clearly and to say exactly what one means. It is unfortunate that most people who write about the relation between genetics and evolution without the intellectual prop of mathematics are either incomprehensible or wrong, and not infrequently both. Dawkins is a happy exception to this rule.

14

Natural Selection of Culture?

The formal similarities between biological evolution and human history have repeatedly tempted students of one topic to borrow ideas from the other. The most famous and fruitful example of borrowing by biologists of an idea from the human sciences was the use made, by both Darwin and Wallace, of Malthus's picture of human competition for resources as a foundation for their own theory of evolution by natural selection. At a less exalted level, I have myself spent much of the last fifteen years applying the mathematical theory of games, first developed for use in economics, to solve problems in evolution. Indeed, I am by no means the only recent biologist to exploit mathematical economics as a source of ideas.

Biologists have, by and large, been eager to borrow ideas from the human sciences. Borrowing in the other direction is less well regarded. The reason for this ill-repute is not far to seek: biological ideas have too often been used, not as potentially valuable research tools, but as a moral justification of policies that might otherwise seem dubious. The Social Darwinists, at the end of the last century, used Darwin's ideas to justify *laissez-faire* capitalism and to oppose economic measures aimed at helping the underprivileged. More recently, the Nazis used biological terminology – they can hardly be said to have used biological ideas – to justify genocide. It would, however, be a great pity of this improper transfer of ideas from biology to the human sciences were to blind us to the possibilities of a fruitful transfer. Boyd and Richerson's book, which offers a Darwinian theory of the evolution of culture – 'the transmission from one generation to the next, via teaching and imitation, of knowledge, values, and other factors that influence behaviour' – is the outcome of much hard and careful thinking. I approached it with a good deal of distrust and trepidation, but am persuaded that they have something of real value to offer.

It may be as well to start by explaining what they do not say. They are not offering a genetical interpretation of society. There are, in fact, two very different kinds of genetical interpretation possible. The first, and less plausible, is the view that the differences between societies are caused by genetic differences between the members of the societies. The most consistent proponent of that view, the late C. D. Darlington, went so far as to argue that language differences are in part caused by genetically determined differences in the vocal tract, and that what is wrong with British agriculture is that the farmers are too inbred.

Boyd and Richerson have little sympathy with this view, even in its less extreme forms. They think that the evidence tends to show that the genetic causes of cognitive and temperamental differences between races are of trivial importance. There is, however, a very different, but still genetical, view. This is that there are universal features of human nature that determine the nature of human societies. There is a sense in which this is obviously true. The reason why human societies are different from the societies of chimpanzees or hunting dogs is that human beings are genetically different. However, Boyd and Richerson argue that the nature of specific human societies cannot be explained merely by the interaction between a universal human nature and the nature of the environment – of soil, climate, and so on. This is so because the behaviour of human beings is determined, not only by their immediate environment, and by the information coded in their genes, but by information that is culturally transmitted. It is the nature and effect of this cultural transmission that they discuss in this book.

They define culture as 'information capable of affecting individuals' phenotypes which they acquire from other conspecifics [i.e., members of the same species] by teaching or imitation.' The critical word here is 'information'. They picture a person's behaviour as being determined by his or her immediate environment, and by two kinds of information, transmitted along two channels, one cultural and one genetic: hence they write of a 'dual inheritance system'. There is nothing about this picture that one could easily object to, and nothing about it that is particularly novel either. What is novel is the way in which they analyse the behaviour of such a system by borrowing concepts, and a style of thought, from evolutionary biology.

Their method is to construct simple mathematical models, aimed at studying the effects of different assumptions that can be made about the way an inheritance system works. Why do they concentrate on simple models? Any simple model of culture will leave out of account many important factors. Should we not, therefore, put into the model everything that we think might be important, even if the price is to end up with a very complex model? The experience of biologists has been that the construction of such complex models is a waste of time, essentially because, as Boyd and Richerson rightly say, 'To substitute an ill-understood model of the world for the ill-understood world is not progress.' It has also been our experience that very simple models can be helpful in answering important questions. To give examples, the questions I have myself tried to answer using simple models include: 'Why are the two sexes different?' 'Why are some species composed of males and females, and others of hermaphrodites?' and 'Why are some hermaphrodites capable of self-fertilisation, and others not?'

There are two reasons why simple mathematical models are helpful in answering such questions. First, in constructing such a model, you are forced to make your assumptions explicit – or, at the very least, it is possible for others to discover what you assumed, even if you were not aware of it. Second, you can find out what is the simplest set of assumptions that will give rise to the phenomenon that you are trying to explain. For these reasons it is increasingly the case, both in ecology and evolution theory, that anyone who claims that he has an explanation of some phenomenon is expected to back up his claim by presenting a formal model. This style Boyd and Richerson adopt in their book. Some readers may find the method discouraging. However, they can follow the main theme of the book while skipping the mathematical bits; the advantage of including the mathematical demonstrations is that, when in doubt, you can find out what they are really saying, and can confirm that their conclusions really do follow from their assumptions.

Their book is not the first attempt to develop formal models of cultural inheritance. Two previous attempts were made by L. L. Cavalli-Sforza and M. W. Feldman, and by C. Lumsden and E. O. Wilson. The second of these is deeply unsatisfactory, for reasons I have discussed elsewhere. Cavalli-Sforza and Feldman, in contrast, did an excellent and scholarly job in a book published in 1981, and their work has been built on by Boyd and Richerson. However, I did find the present book more rewarding, for a number of reasons. Their review of empirical data from the social sciences, both in support of some of the assumptions of their models, and to illustrate the phenomena that their models might help to explain, is much richer than in the earlier book. They analyse a wider range of models, and so reach more conclusions, and in particular more unexpected and unobvious conclusions. They make a greater effort to place their own work in the context of what others have done. Finally, they tackle a problem that is of particular interest to me as an evolutionist: How can we explain the genetic evolution of those human characteristics that are needed if extensive cultural inheritance is to be possible?

In addition to borrowing a style of analysis from biology, they have also borrowed the central concept of natural selection. As was clear to Darwin, selective survival and reproduction will cause change, but only if the traits that influence survival are inherited. If cultural traits are inherited, by teaching and imitation, and if different traits have different likelihoods of being transmitted, then the Darwinian model of evolution by natural selection applies, even though the inheritance is not mediated by DNA.

To see what this view implies in more detail, and, incidentally, to see what Boyd and Richerson have in mind when they speak of simple models, consider the two following possibilities. First, suppose that a child acquires some particular cultural trait from one or other of its parents, with equal likelihood, such as a tendency to be altruistic. Suppose further that some variant of the trait causes its possessors to have, on average, more children than they would have if they had not acquired the trait. Then that variant will become more common, by a process exactly analogous to the spread of a genetically determined trait by natural selection.

Now consider an alternative model, in which differences in biological fecundity play no role. Suppose that individuals acquire some trait by imitating, not their parents, but some other member of their social group. They tend to imitate those members who are 'successful' according to some criterion: for example, to offer an implausible fantasy, they imitate those professors who attract more students to their lectures. Then, if some cultural variant leads to success (i.e. to attracting large audiences), that variant will tend to become common in the population.

I have deliberately chosen two of the simplest possible models of the natural selection of cultural traits, so as to bring out the gist of the authors' argument. When they speak of the natural selection of cultural traits, they think that such selection brings about evolutionary change because those traits are culturally inherited through imitation and teaching: No genetic change need occur. They recognise, however, that there are two important ways in which the analogy between cultural and biological evolution breaks down. In biology, mutations are nonadaptive – i.e. they do not, in most cases, increase the fitness of the organism – and acquired characters are not inherited: In cultural inheritance, the analogous statements may not be true. Boyd and Richerson seem to give little importance to the faculty of reason. They do accept that a person is more likely to acquire a new habit if he thinks it will pay him to do so, although they argue that the connections between human actions and their consequences are often so hard to calculate that we tend to rely on custom rather than reason. All the same, the existence of human reason does mean that cultural innovation, unlike mutation, can be adaptive. The need to allow for the inheritance of acquired characters is even more obvious: if I learn something I can tell my children and have some hopes they too will learn it.

My own suspicion is that these structural differences between culture and genetics will inevitably limit the usefulness of the kind of theory presented in this book. The explanatory power of evolutionary theory rests largely on three assumptions: that mutation is non-adaptive, that acquired characters are not inherited, and that inheritance is Mendelian – that is, it is atomic, and we inherit the atoms, or genes, equally from our two parents, and from no one else. In the cultural analogy, none of these things is true. This must severely limit the ability of a theory of cultural inheritance to say what can happen and, more importantly, what cannot happen.

I conclude this review by considering one particular model in some detail. In a chapter which opens with an account of the Japanese kamikaze pilots in the last world war, the authors develop a theory to account for the tendency of human beings to behave altruistically towards the groups to which they belong by making sacrifices for the common good. In their view, the beliefs that led the kamikaze pilots to die for their country are an extreme example of such tendencies. Biologists have spent much time on the analogous phenomenon in animals: We have explanations – such as the fact that the altruist may share genes with the recipient of its altruism, and it is genes, not individuals, that matter in evolution – but they are ones that work only for altruistic behaviour among the members of small groups. Boyd and Richerson start by analysing a model in which individuals acquire cultural traits by copying those that are most common in the population. Of course, if I copy the first person I meet, then I am more likely to acquire a common than a rare trait, but Boyd and Richerson are here assuming a more

extreme form of imitation in which I adopt whichever trait
I observe to be the most common, after observing many
people. They show that such a 'conformist' habit of acquiring
culture can be an advantage to the individual in a variable
environment, in which different traits are favoured in
different places. This, they suggest, has led to a tendency
to adopt, when in doubt, whatever are the common beliefs
and customs in the society to which one belongs. I am not
clear whether they think that this conformist tendency is
genetically programmed, or itself something that evolves
culturally.

So far, there is nothing in the argument to imply that
these customs should be self-sacrificing. At this point,
therefore, they introduce the concept of 'cultural group
selection'. Given conformist individuals, they show that
cultural differences between groups will be exaggerated. If
the cultural practices of different groups differ in their
effectiveness in ensuring the long-term success of the group,
then group selection can lead to the evolution of cooperative
and self-sacrificing practices. For example, note that the
argument is that cultural, not genetic, evolution will occur.
All that is perhaps genetic is the conformist habit in acquiring
culture.

The argument is ingenious, and I think it is logically
sound, but I am not convinced that it is either necessary or
sufficient for explaining the evolution of cultural traits. First,
is it necessary? Boyd and Richerson argue that it is hard to
account for altruistic behaviour in large groups if behaviour
is rational, even if everyone would benefit if everyone else
was altruistic. Essentially their reason is that it is hard to
explain, on purely rational grounds, why individuals in a
cooperative society do not take the benefits of cooperation
without paying the full costs. I do not see their difficulty.
In the society I inhabit, public goods like roads, services,
and even a health service are paid for by taxes. If I do not
pay my taxes, I am punished, so it is rational to pay taxes.
If I am asked, as I am every five years or so, whether I
prefer that everyone should pay taxes and should have access
to public goods, or that taxes should be reduced and the
public goods disappear, it is rational for me to vote for the

former. Of course, looking at the world as it is, no one could suppose that reason always leads to beneficial cooperation. The breakdown of rational cooperation arises, I think, because of the existence within society of different groups with different interests.

The difficulty for any account of society that assumes that individuals behave rationally is partly that experimental psychologists find little support for such an optimistic view. For example, people form conclusions on the basis of insufficient data, and are then reluctant to change their minds when confronted by disconfirming evidence. It is also partly that the world is full of groups whose behaviour benefits neither the group as a whole nor the individuals that compose it. The activities of most terrorists are unlikely to benefit either the individual terrorist, or the group to which he belongs, or the cause for which that group claims to be fighting. Such groups are often motivated by fanatical beliefs which, to outsiders, seem irrational and self-destructive. The argument developed by Boyd and Richerson perhaps does something to explain such fanaticism, but I wonder whether it is sufficient to explain it. I do not have anything better to offer. The capacity of our species, for good or ill, to be swayed by myths I find a continuing and as yet unanswered puzzle.

I found this book much more fun to read than I had expected. It can be read without following all the mathematics in it, but as I have noted it is important to have the mathematics there, so that one can on occasion find out exactly what Boyd and Richerson mean, and can satisfy oneself that their assumptions really do lead to their conclusions. They have made a brave attempt to integrate their abstract models with data from psychology and anthropology. They have helped to distinguish between and to evaluate different theories of sociobiology and of culture. They have proposed a number of ingenious models to account for the behaviour of the oddest species that has yet evolved.

PART 3

DID DARWIN GET IT RIGHT?

The latest attempt to dethrone Darwinism as the central theory of biology is the 'Punctuationist' theory of Gould, Eldredge and Stanley. It has some positive features, most notably, as I argue in *Palaeontology at the High Table*, that it emphasises the role that palaeontology can play in evolutionary theory. Nevertheless, I believe it to be largely mistaken, for reasons I explain at some length in *Current Controversies in Evolutionary Biology*, and more briefly, for a less professional audience, in *Did Darwin get it Right?*

There have always been schools critical of Darwin, partly because there is always fame to be won in science by killing the king and partly for the ideological reasons I discuss in *Science, Ideology and Myth*. I have little sympathy, however, for schools that have been constructed by bringing together everyone who has something anti-Darwinian to say, however mutually contradictory their own views may be. This may explain the rather ill-tempered tone of *Do we need a New Evolutionary Paradigm?*

15

Palaeontology at the High Table

The Tanner lectures, recently delivered by Stephen J. Gould of Harvard University, and followed by a panel discussion, provide an opportunity to assess the current contribution of palaeontology to evolutionary theory. It might be supposed that this contribution would be crucial, but, at least until recently, that has not been so. The palaeontologist G. G. Simpson was one of the main architects of the 'modern synthesis' that emerged in the 1940s, but his role was to show that the facts of palaeontology were consistent with the mechanisms of natural selection and geographical speciation proposed by the neontologists (a term used by palaeontologists to describe the rest of us), rather than to propose novel mechanisms of his own. Since that time, the attitude of population geneticists to any palaeontologist rash enough to offer a contribution to evolutionary theory has been to tell him to go away and find another fossil, and not to bother the grownups.

In the last ten years, however, this situation has been changed by the work of a group of palaeontologists, of whom Gould has been a leading figure. In his lectures, he stressed two theoretical modifications to the Darwinian scheme – 'punctuated equilibria' and a hierarchical view of evolution. Punctuational theory holds that most evolutionary change occurs when lineages split – that is, when a single species gives rise to two. Between such events, species remain morphologically unchanged, often for millions of years: they remain in 'stasis'. In itself, the theory says nothing about the mechanism of the changes when they do occur, other than that they are rapid in geological terms: they could still be slow to a geneticist, and brought about by the familiar process of natural selection. As J. S. Jones (University College, London) remarked in an earlier discussion, one man's punctuation is another man's gradualism.

Little was said in Cambridge about the empirical evidence for punctuation. Gould treated it as a fact the whole world knows: as something proved by Aunt Jobisca's theorem. As an outsider, I am persuaded that morphological evolution proceeds at very different rates at different times, but can see little evidence that rapid change is necessarily associated with the splitting of lineages (although one would expect this often to be the case, when part of a species enters a new ecological niche). What is new is the emphasis on stasis. As David Wake (University of California, Berkeley) pointed out during the discussion, the existence today of pairs of species which are morphologically indistinguishable, or almost so, but whose proteins have diverged sufficiently to suggest that they have been separate for millions of years, supports the reality of stasis. Stasis, then, is a phenomenon that calls for an explanation.

Geneticists and palaeontologists tend to differ in the explanations they prefer. Population geneticists argue that if a species does not change, this is because it is subject to 'normalising selection' – that is, selection that eliminates the extreme phenotypes and favours the norm. They have two main reasons for this belief. First, if directional selection is applied to almost any characteristic, in nature or in the laboratory, the population will change. Second, most species with a wide geographical range do vary in structure and behaviour, sometimes to such an extent that populations from different regions would be regarded as different species if they were not connected by a series of intermediate forms. (In case anyone supposes that membership of a species is decided unambiguously by hybridisation, it is the case that different species often produce fertile hybrids, and populations of the same 'species' often prove, on investigation, to produce sterile or inviable hybrids.)

In reply, Gould asks how it could possibly be the case that normalising selection could favour the same phenotypes over long periods, during which the environment may change drastically. Instead, he proposes a concept of species stability arising from developmental constraints. The constancy of a species would thus resemble the constancy of a chemical element: it would represent a stable state of matter, which can be changed only by an unusual event. C. H. Waddington's idea of 'canalisation' has been quoted in support of this view. Waddington did indeed point out that the typical 'wild-type' morphology of a species is usually well buffered against disturbance during development, whereas the phenotypes of individuals with a given mutation are highly variable between individuals. However, for Waddington, the uniformity of the wild type was itself the product of normalising selection, and not a manifestation of an intrinsically stable state.

Hierarchies

The debate about stasis and punctuation will no doubt continue. Gould's second thesis concerned the hierarchical structure of evolution. Darwin, he argued, thought of natural selection as favouring some individual organisms at the expense of others. However, selection, and also chance, can be effective at two other levels, namely the gene and the species. At the level of the gene, we now know that there are processes (gene conversion, unequal crossing-over, transposition) that enable genes to multiply out of phase with the organisms in which they find themselves. It follows that a gene can increase in frequency by virtue of its ability to multiply horizontally within an individual, and not only by virtue of its effects on the individual's phenotype.

Two points were made about this level of selection during the discussion. John Fincham (Cambridge University) pointed out that since transposons are in effect parasites, we can expect them to evolve 'self-restraint', as some parasites do: it does not pay to kill the goose that will transmit you to future generations. A more fundamental disagreement was raised by Gabriel Dover's (Cambridge University) claims for 'molecular drive' – a portmanteau term for the three processes listed above. I may give an unfair account of this disagreement, because I was one of the protagonists. Dover asserted that there is nothing that natural selection can do that cannot be done by molecular drive. I argued in reply that this is to overlook the fact that natural selection can generate highly improbable adaptations at the organismic level, whereas molecular drive cannot. As I see it, Dover's view would throw out the Darwinian baby with the bathwater.

Species selection

Palaeontologists can contribute little to this debate, but have a lot to say about the level of species selection. The concepts can best be explained by some examples. First, during the Cretaceous, the Volute gastropods gradually replaced their competitors. A plausible explanation is that Volute development did not involve a stage of mobile planktonic larvae, whereas the development of their competitors did. As a result, geographical speciation was more frequent among Volutes, and the number of species increased. In effect, Volute species had a higher 'birth rate'. If this explanation is correct, it applies the concept of natural selection, but replaces individuals by species as the units selected.

Now consider a second example, the mammalian secondary palate. This is a structure that enables a mammal to breathe and chew at the same time. Gould would agree that such a structure (together with associated changes in teeth, jaw muscles and articulation) could arise only by natural selection at the level of individuals. Suppose, however, that mammalian species came into competition with reptiles lacking these adaptations. If the mammals outcompeted the reptiles because they could chew better, then mammalian species would replace reptilian ones. However, Gould would not call this species selection, because the success of the mammals would depend on something individual animals do (chew) and not on something species do: in contrast, in the Volute example, speciation is something species and not individuals do.

There is a third scenario. For example, at the end of the Cretaceous the Dinosaurs became extinct, whereas the mammals survived and radiated. I do not know why this was so, but let us suppose that the mammals survived because they are homoiotherms, and that their ability to chew had nothing to do with it. It would still be true that the Cretaceous extinctions led to an increase in the proportion of species with secondary palates, but not because of any selection for the palate itself. This is what Elizabeth Vrba has called the 'effect hypothesis'. It is an analogue, at the level of organs, of the process of 'hitch-hiking' at the level of genes.

The hierarchical view of evolution, then, is that processes of selection and stochastic drift go on at the level of genes and of species, as well as of individuals. As an old-fashioned proponent of the modern synthesis, I have no difficulty in accepting this, so long as no one expects me to believe that adaptations of individuals can be explained by molecular drive, and so long as the concept of species selection is confined to qualities, such as speciation rate and evolutionary rate, that are properties of species and not of individuals. However, there remains plenty of room for disagreement about the relative importance of these various levels, or – what amounts to the same thing – about the relative effectiveness of selection at different levels.

In from the cold

This brings me to what I see as the greatest impact that palaeontology is having on the way we see the mechanisms of evolution. We have been familiar for a long time with the dramatic disappearance of the Dinosaurs at the end of the Cretaceous. It is now apparent that massive extinctions, involving many different taxa, have been a repeated feature of evolution. Adolf Seilacher (Tubingen University) and Anthony Hallam (Birmingham University), the two palaeontologists on the panel, agreed with Gould on this, although there was disagreement about whether these events have been periodic or irregular, and whether they are caused by extraterrestrial events (meteorites, asteroids) or by terrestrial ones (Hallam, for example, emphasises the role of changes in the area of the continental shelf caused by continental drift – see *Nature* 308, 686; 1984). The impact of these extinctions is not random; in any given event, some taxa are more affected than others: Seilacher stressed the need for a quantitative study of this, to replace the somewhat anecdotal picture we now have.

In addition to the problem of their causation (which at present seems to be a problem for geologists and astronomers, although Seilacher did not rule out the possibility that some extinctions had biotic causes), these extinctions raise questions for evolutionary biologists. Is it possible that evolutionary change would slow down and stop in the absence of changes in the physical environment? As Manfred Eigen has pointed out, the simplest evolving systems (populations of RNA molecules in test tubes) reach a global optimum and then stop. Are extinctions, then, a necessary motive force of evolution? A second question concerns the relation between extinction and radiation. Ecologists tend to see nature as dominated by competition. They would therefore expect the extinction of one species, or group of species, to be caused by competition from another taxon. Most palaeontologists read the fossil record differently. The Dinosaurs, they believe, became extinct for reasons that had little to do with competition from the mammals. Only subsequently did the mammals, which had been around for as long as the Dinosaurs, radiate to fill the empty space. The same general pattern, they think, has held for other major taxonomic replacements. Not all palaeontologists would agree, but I think this is the majority view. I find it surprising: I would have expected a major cause of extinction to be competition from other taxa.

The Tanner lectures were an entertaining and stimulating occasion. The palaeontologists have too long been missing from the high table. Welcome back.

16

Current Controversies in Evolutionary Biology

Evolutionary biologists are arguing about many things – how and why sex evolved, whether some DNA is 'selfish', how eukaryotes arose, why some animals live socially, and so on. These problems are, in the main, debated within the shared assumptions of 'neo-Darwinism' or 'the modern synthesis'. Recently, however, a group of palaeontologists, of whom Gould, Eldredge and Stanley have been the most prominent, have announced that the modern synthesis is soon to be swept away, to be replaced by the new paradigm of stasis and punctuation.

In science, a theory is not abandoned unless an alternative theory already exists, ready to replace it. My object in this essay is to identify this alternative, and to explain why I do not find it particularly persuasive. The main aim will be to clarify ideas.

The punctuationist position consists of a minor and a major claim. The minor claim is that the typical pattern of the evolution of species, as revealed by the fossil record, is one of long periods of stasis during which little significant change occurs, interrupted by brief periods of rapid change associated with the splitting of species into two. The major claim is that it is a consequence of this observation, together

with a study of development, that the large-scale features of
evolution are not the result of the accumulation of changes
occurring in populations because of natural selection, together
with the processes of speciation as understood by the
proponents of the modern synthesis. In brief, macro-
evolution can be uncoupled from micro-evolution.

I shall say little about the minor claim. Ultimately, it is
a matter of empirical test. I have only two comments to
make. The first is that the answer may be hard to come by,
as is so often the case in evolutionary biology – compare,
for example, the continuing debate about the neutral mutation
theory. One difficulty is that a sudden transition from form
A to form B in one place may (or may not) conceal the fact
that A changed gradually to B in some other place, and that
B subsequently migrated to replace A in the area studied. It
may, therefore, prove to be easier to establish (or to dismiss)
the reality of stasis than to study the nature of the transitions.
My second comment is that it will be of little use to analyse
the durations in the fossil record of particular named forms,
as Stanley attempts, because this is to study the habits of
taxonomists rather than the evolution of organisms. There
is no alternative to a statistical study of populations.

At present, I am unconvinced that stasis and punctuation
are typical, although I am satisfied that they occur and was
persuaded long ago by Simpson that evolution can proceed
at very variable rates. The problem for a population geneticist
concerns stasis rather than change. A change comparable to
that between species which was completed in 1000 generations
would be rapid to a palaeontologist but slow to a population
geneticist: consider the changes produced by artifical selection
in dogs in little more than 1000 generations.

For stasis, two explanations have been proposed. One is
that the form of a species cannot readily be changed by
selection because of constraints during development. This
idea lies at the core of the new theory, and I will, therefore,
postpone discussion of it until later. The other is that a
species does not change because of stabilizing selection; that
is, changed individuals are of lower fitness. This raises the
question of why the selective optimum should remain
unchanged for millions of years. It certainly seems incompat-

ible with Van Valen's 'Red Queen' picture of evolution, according to which each species is evolving as fast as it can to adapt to changes in the others. Nils Stenseth and I have been trying for some time to formulate Van Valen's ideas more precisely. It seems to us that a model combining both ecological and evolutionary time-scales can lead to one of two pictures, between which only the fossil record can decide. One is the Red Queen hypothesis, with evolution, extinction and speciation continuing even in a uniform physical environment. The other is one in which evolutionary change gradually slows down and stops in the absence of changes in the physical environment. We, therefore, have some interest in the outcome of the argument about stasis in the fossil record.

Before leaving the question of stasis, there is one other crucial point to be discussed. This is the difficulty of reconciling the assertion that species display stasis in time with the fact that they manifestly do not do so in space. The essential point can be made by a quotation from Mayr: 'We find that in every actively evolving genus there are populations that are hardly different from each other, others that are as different as subspecies, others that have almost reached species level, and finally still others that are full species.' In this continuum of degrees of difference, to what is the punctuational event supposed to correspond?

At this point, I cannot resist digressing for a moment to discuss a curious twist in the micro-history of science. In the debate about macro-evolution, an alliance has been formed between punctuationalists and 'cladists' – the followers of a taxonomic methodology first proposed by Hennig. There is, certainly, no necessary connection between the two points of view; Gould, for example, has said that he is not a cladist. For some time I was unable to see any link between the two positions. Hennig himself was almost too committed to the modern synthesis. However, a recent reading of Hennig, and of Hull, has suggested a connection.

Hennig was concerned that a hierarchical classification (into species, genera, families, etc.) should accurately reflect phylogeny. He went further, and insisted that a classification should reveal to anyone looking at it the phylogenetic hypothesis held by the classifier. If all the species classified are contemporary, this raises no difficulty of principle. If some species are possible ancestors of others, ambiguities arise. Hennig pointed out that these ambiguities are overcome if certain rules are obeyed when naming fossils. The essential rules are that a lineage must *not* change its name except when it splits into two, and that when it does split, both daughter species *must* be given new names. Thus a lineage that changes without splitting can have only one name, no matter how great the change; a species that 'buds off' another without itself changing must, nevertheless, change its name.

For Hennig, these were rules of convenience, needed if classifications were to yield unambiguous phylogenies. For punctuationists, they have become assertions about the world – species do not change except when they split. An odd reason for an alliance! Matters have since become still more confused. Hennig's central belief was that classification should reflect phylogeny. His methods – in particular, the distinction between 'plesiomorphic' (primitive) and 'apomorphic' (derived) traits – make sense only if one supposes that the objects being classified have arisen by a process of branching evolution. Yet there exists today a school of lapsed cladists who continue to use Hennig's methods, while denying that evolution has any bearing on taxonomy. It is as if a lapsed Catholic were to continue to attend mass.

This discussion of cladism has been a digression: I now turn to the major claim, the so-called uncoupling of macro-evolution from the processes of selection in populations. This is best understood in terms of two proposals – 'hopeful monsters' and 'species selection' – which have so far been offered as alternative mechanisms of evolutionary change.

The idea of 'hopeful monsters' goes back forty years to Goldschmidt. Recently, Gould and others have suggested that his ideas, although rejected at the time, may be due for a revival. I want to argue that there are two rather different concepts involved here – which I shall refer to as 'hopeful monsters' and 'systemic mutations' – of which only the latter is clearly incompatible with the modern synthesis. By a 'hopeful monster' I shall mean here no more than an individual carrying a genetic mutation of large phenotypic effect. For the term 'systemic mutation' I accept Goldschmidt's meaning of a complete repatterning of the genome, by means of a number of rearrangements of the chromosomal material, giving rise in a single step to a new species or higher taxonomic group. It is this latter concept which was, in my view, rightly rejected; I can see no reason for wishing to revive it. In effect, it accounts for novelty by postulating a miracle.

What of a hopeful monster in the simpler sense of a mutant individual so different from typical members of the species that it is able to adopt some new habit or survive in some new environment? Darwinists have had two rather different reasons for expecting evolution to proceed by small steps. The first concerns the 'perfection of animals' (Cain). Even if not perfect, the degree of adaptive fit between organisms and their ways of life can be very striking. A detailed adaptive fit, if produced by natural selection, requires that there is a large amount of finely-graded variation for selection to act on. I believe this is correct, but it does not follow that *all* changes should be small – it does not rule out hopeful monsters. It means only that a hopeful monster will need further fine tuning by selection of smaller variants before its descendants achieve detailed adaptation to the new way of life.

A second argument for gradual change, owed to Fisher, would rule out hopeful monsters if it were accepted. Fisher argued that existing organisms are close to a selective optimum and hence that a large change is less likely to improve adaptation than a small one. A large, effectively random change in a complex integrated mechanism is almost certain to have disastrous consequences. This may not be as true of organisms as it is, say, of motorcars. In a paper read at the ICSEB at Vancouver in 1980, Thomson argued that development is so buffered that a single, large change may be compensated for by many secondary changes in the ontogeny of a single individual, without waiting for further genetic change. To illustrate his point, he referred to a goat described by Slijper that was born without front legs and consequently adopted a bipedal gait. The anatomy of the backbone, the orientations and lengths of its processes, and the sizes and insertions of muscles and tendons, all changed in ways that could be interpreted as adaptations to a bipedal gait. Presumably, these changes were brought about because bone grows along lines of compression, tendons along lines of tension, muscles hypertrophy if they are used, and so on.

For these kinds of reason, Thomson argued that a monster may occasionally be hopeful and survive to become the ancestor of organisms with new ways of life. As it happens, I agree with him, at least this far. Indeed, the argument just outlined – including, oddly enough, the use of Slijper's goat to illustrate the point – was included in my book, *The Theory of Evolution*, published in 1958. That book was written to explain the ideas of the modern synthesis to a non-professional readership. I did not see then, and do not see now, any contradiction between neo-Darwinism and the idea of hopeful monsters, at least in the sense of a mutant of large phenotypic effect. The essential point is that the fate of hopeful monsters, like that of other mutants, depends on the operation of natural selection in populations.

The preceding discussion, however, is hypothetical. It amounts to saying that one cannot rule out hopeful monsters *a priori*. But do they happen? As Lande pointed out at the Chicago macro-evolution meeting in 1980, the genetic evidence suggests that the differences between morphologically distinct species and varieties is polygenic and does not involve one mutation of large effect. The conclusion is based on an analysis of F_1, F_2 and backcross hybrids; more studies of this kind would be welcome. We are now familiar with the idea that gradual changes in the parameters of a dynamic system can, at critical points, lead to sudden and discontinuous changes in system behaviour. It seems certain that gradual changes in genetic constitution can lead to discontinuous changes in phenotype. The only question at issue is how often large changes have contributed to evolutionary change.

The occasional incorporation in evolution of mutations with large morphological effects is an interesting, if unproved, hypothesis. Even if true, it would hardly qualify as a paradigm shift. I now turn to the second theoretical proposal of the punctuationists – 'species selection'. The idea is as follows. When a new species arises by splitting, it acquires new characteristics. The direction of change is assumed to be random relative to any large-scale evolutionary trends and not to be caused by within-population selection. (This is known as 'Wright's rule' in accordance with Wright's earlier views on the random nature of speciation.) However, species will differ in their likelihood of extinction, or of further splitting, and these likelihoods will depend on their characteristics. Hence selection between species will cause trends in species characteristics.

This theory is isomorphic with evolution in a population of parthenogens, with speciation replacing birth, extinction replacing death, and the acquisition of new, random characteristics at the moment of speciation replacing mutation. Logically, some such process must occur – although there is no reason to assume the truth of 'Wright's rule'. I have argued elsewhere that species selection is relevant to the maintenance of sexual reproduction, although not necessarily the only, or even the most important, process involved. But is it responsible for macro-evolutionary trends?

There is one category of evolutionary event that must often have occurred, and which might be taken as an example of species selection, but, in my view, misleadingly so. This is the competitive replacement of one group of species by another; examples are the replacement of the multituberculates by the rodents, or of the creodonts by the modern carnivores. In both cases, it is at least plausible that the replacement occurred because individuals of the new taxon were competitively superior to individuals of the old one. If so, should one call it 'species selection'? In a sense, this is a purely semantic issue. However, the importance of replacements of this kind has long been recognised and is not a contribution of the punctuationists. Furthermore, such replacements in no way uncouple macro-evolution from micro-evolution, provided, of course, that the characteristics by virtue of which, for example, the rodents replaced the multituberculates arose in the first instance by natural selection. Hence, taxonomic replacements of this kind, however important, cannot in the present context be taken as examples of species selection.

The difficulty in regarding species selection as an important evolutionary force can be seen best by discussing a concrete example. Consider the trends in the evolution of the mammal-like reptiles. In the skull, there were a group of changes associated with the evolution of chewing – reduction of tooth replacement to one, differentiation of the teeth, opening up of the dermal roof of the skull, and the development of a bony secondary palate. The adaptive significance of the reduction in the number of bones in the lower jaw, the acquisition of a new jaw articulation, and the incorporation of quadrate and articular into the middle ear, is less clear to me. Postcranially, there is a whole series of changes in the backbone, limb girdles and limbs associated with a gait that involves flexing the backbone in a vertical rather than a horizontal plane (referring back to Slijper's goat, many of these changes are not of a kind that could have have arisen as secondary ontogenetic responses to a single primary change). Changes in the soft parts, associated with homoiothermy, viviparity, a double circulation, and so forth, are not recorded in the fossil record, but were presumably occurring.

I find it hard to believe that anyone seriously thinks that these changes were the result of species selection. There are two difficulties, one quantitative and the other logical. The quantitative difficulty is that the amount of change that can be produced by selection is limited by the number of births and deaths (i.e. speciations and extinctions) that occur. I cannot be precise about this, because I do not know how many species extinctions occurred, or how many independent changes in character states were incorporated, but I doubt whether a quantitatively plausible account could be given.

The graver difficulty, however, is the logical one. Species do not chew or gallop or keep warm or bear their young alive: individual animals do these things. Hence, if I am right in thinking that the secondary palate evolved because it enables an animal to chew and breathe at the same time, there is no way in which 'species selection' could be responsible for its evolution, *except* in so far as species survive if and only if the individuals who compose them survive.

The point may become clearer if I revert for a moment to the evolution of sex. Species evolve, but individual animals do not. Hence sexual reproduction affects a property of a species – namely, the capacity to evolve. It is, therefore, reasonable to speak of sexual reproduction evolving by species selection (as R. A. Fisher in effect did: whether or not he was right is still a matter of debate, but at least his argument was not illogical). The claim would be correct if:

1. Mutations from sexual to asexual reproduction are commoner than the reverse, which is almost certainly true.
2. Within populations, asexual reproduction tends to replace sexual.
3. Sexually reproducing species evolve faster and are, therefore, less likely to become extinct than parthenogenetic populations.

In the case of the secondary palate, however, the claim that it evolved by species selection would have to mean something like the following. There existed a species, A, all of whose members had a partially developed palate. There arose from A a second species, B, reproductively isolated from A, all of whose members had a better developed palate, and a third species, C, whose members had a less well-developed palate. In competition, the B individuals outcompeted the A and C individuals, because they could chew better, so that only species B survived.

Note that, in this scheme, the reproductive isolation and the changed palate would have to arise simultaneously. If all that happened was that, in the original species A, some individuals arose with better palates, we are back with good old-fashioned natural selection in populations.

Would it make any difference if we abandoned my assumption that a secondary palate evolved because it enables animals to chew and breathe simultaneously? Maybe the palate influences individual survival or reproduction in some other way, in which case the essence of my argument is unchanged. Perhaps it does not affect survival or reproduction at all, but is associated with something that does; if so, my argument is still unchanged. Finally, and implausibly, it may be that a secondary palate does not affect survival and is not correlated with anything that does. If so, the species selection argument collapses altogether, because, if the palate affects species survival at all, it must do so via some effect on the individual that has the palate. If there are no such effects, we are left, to coin a term, with 'species drift'. To explain the major features of evolution by species drift would surely be the ultimate absurdity.

In the light of this discussion, we can now recognise four different processes which might be called 'species selection':

1. Selection operating on emergent properties of the species itself, affecting chances of extinction and/or speciation. Examples are capacity to evolve rapidly (influenced, for example, by sexual reproduction and level of genetic recombination), and likelihood of speciation (influenced, for example, by dispersal behaviour). The use of the term here seems appropriate. However, most evolutionary trends (e.g. in the secondary palate) could not be explained in this way. Even traits that could, in principle, be influenced by this type of species selection may, in practice, be largely determined by individual selection. For example, recombination rate is almost certainly the result of individual selection; the essential point is that there is extensive intraspecific variation in the trait. Dispersal patterns may have been influenced by the individually disadvantageous effects of inbreeding.

2. Selection acting on traits affecting individual survival and/or reproduction, but that arise suddenly at the time of speciation, simultaneously with reproductive isolation, and not as the result of within-population selection. If such a process occurs, I have no objection to calling it species selection. Even though the relevant traits are properties of individuals and not of species, the adaptedness of such traits would arise because of the survival and splitting of some species and the extinction of others. My objection to invoking this type of species selection is that I do not think species arise in this way (except in speciation by polyploidy); if they

did, I do not think the number of speciation and extinction events would be large enough to account for the extent of adaptation actually observed.

3. The replacement of one species, or group of species, by another, because individuals of the successful species have selectively superior traits that evolved, in the first place, by individual selection within a population (for example, the replacement of multituberculates by rodents, supposing that the traits responsible for the success of the rodents did arise in the first instance by within-population selection). In my view, it would be misleading to describe such events as cases of species selection.

4. The radiation of a taxon into a new adaptive zone by virtue of traits that evolved, in the first instance, by within-population selection. Again, as for category (3), I think it would be misleading to use the term species selection.

It is important that proponents of species selection should make clear which, if any, of these meanings they intend. Discussions at the conference with Gould suggest that he intends meaning (1), and perhaps (4), but not (2) or (3); however, he will no doubt clarify his own position. I suspect that Stanley intends meaning (2), because it corresponds to his view of the nature of speciation.

If, as seems to me appropriate, species selection is confined to meanings (1) and (2), it follows that (with the exception of a few traits, such as sexual reproduction or powers of dispersal, which confer properties on the species as such as well as on the individuals that compose it) the concept applies only if new species arise suddenly, with new characteristics, and are at once isolated reproductively from the ancestral species. Only if this is true does the concept have any meaning.

The hypothesis of species selection, therefore, arises from a typological view of species: The view that only certain forms are possible, and that intermediates are ruled out. There is nothing ridiculous *a priori* about this view. A typological view of chemistry would be correct. The laws of physics permit the existence of only certain kinds of atom; there are no intermediates between helium and hydrogen. What are the equivalent laws that are supposed to be responsible for the fixity of species? Two different but related proposals have been made; that there are 'developmental constraints', and that evolutionary change can occur only in small isolated populations. I shall discuss these points in turn.

The concept of a developmental constraint is best explained by an example. Raup has shown that the shapes of gastropod shells can be described by a single mathematical function, with three parameters which can vary. This is a consequence of the way the shell grows (e.g. that it grows by addition at an edge and not by intercalation). In other words, because of the way the shell grows, only certain shapes are possible; there cannot be gastropods with square shells.

That developmental constraints exist is not at issue. But they cannot by themselves account for stasis, or typological species. They limit the *kinds* of change that can occur, but they do not rule out all change. An infinite variety of different shapes is consistent with Raup's rules, and any one could be changed into any other by minute degrees. Of course, developmental constraints may permit a discrete set of possibilities, rather than an infinite gradation. However, such quantised variation occurs within species, without involving reproductive isolation, and there seems to be no difference between the genetic basis of continuous and quantised variation.

Thus developmental constraints limit the kinds of variation that can arise in a given species, but they do not rule out all variation. I suggested earlier that stabilising selection maintains stasis; it will do so subject to the range of variation that arises. However, developmental constraints alone could not account for stasis, unless it were true that *no* variation is possible, and that is manifestly not so.

It is worth adding that the concept of developmental constraints is by no means new. As an example, my colleague Dr. Brian Charlesworth has drawn my attention to the following remark by H. J. Muller:

The organism cannot be considered as infinitely plastic and certainly not as being equally plastic in all directions, since the directions which the effects of mutations can take are, of course, conditioned by the entire developmental and physiological system resulting from the action of all the other genes already present.

Of developmental constraints, then, one can say that they will inhibit or prevent changes in some directions, but not in all directions. What of the genetic barriers to evolutionary change? It is sometimes asserted that there is some kind of inertia that prevents evolutionary change in large populations. At least as far as our present understanding goes, this assertion is false. It has been supported by references to Lerner's concept of 'genetic homeostasis'; this is an interesting idea, based on the empirical fact that a small population exposed to strong directional selection often reverts partway to its original phenotype when selection is relaxed. Lerner would have been astonished to hear his idea used to support the view that large populations cannot evolve.

There are two points at issue: Can a large population change, and can it split into two without geographical isolation? On the first, I can think of no sensible reason why it cannot, and good reasons why it should be able to sustain evolutionary change for longer than a small population. The second question is more controversial. I shall not pursue the controversy here, because it is not essential to the argument about species selection.

The idea that large populations cannot evolve seems to be simply false. However, there is one *kind* of change that cannot take place in a large population, but may, with low probability, occur in a small one. This is the passage from one adaptive peak to another through a selectively inferior intermediate. The simplest case is the passage, in a diploid, from a population homozygous for one allele to one homozygous for the other, when the heterozygote is of inferior fitness. In the present context, however, the more interesting cases arise when there are epistatic fitness interactions between loci. The simplest case, in a haploid, is the transition from ab to AB when aB and Ab are both of low fitness.

Such events are the subject of one of the oldest controversies in population genetics. Wright thinks that such transitions were important, and Fisher thought that they were not. As a student of Haldane's I can take an impartial view.

It may help at the outset to explain what is *not* at issue. It is not at issue that there are epistatic effects on fitness. Nor is it at issue that the genes present at different loci in a sexually reproducing population are 'co-adapted'. This is self-evidently true – no one supposes that a genome consisting partly of mouse genes and partly of rabbit genes would give rise to a viable animal. The difference of origins need not be so extreme, as shown by the fact that hybrids, in the first or later generations, between closely related species, or even varieties of the same species, often show lowered viability and/or fertility.

The essential point is to understand that the facts of co-adaptation, and of hybrid inviability and infertility, are *not* evidence that populations have in the past crossed adaptive valleys. I show, in Figure 16.1, two simple cases in which a population can evolve from one state to another by a series of favourable gene substitutions, and yet the F_1 hybrids between the terminal states are of lower fitness. I show the simplest cases for which this can be true; as more loci are involved, the likelihood that two populations that have undergone independent evolutionary changes will give inviable hybrids increases.

The reason why the Fisher–Wright debate is so difficult to settle should now be clear; we do not know whether all evolutionary change is a hill-climbing process which could occur in a large population, or whether, occasionally, adaptive valleys are crossed. Perhaps the best reason for thinking that valleys are sometimes crossed is that some kinds of structural changes in chromosomes do seem to imply intermediates of lower fertility.

If, provisionally, we suppose that valley-crossing in small populations has been important in evolution, there can still be differences of opinion about its role. Wright imagined a species divided into a large number of demes with little gene flow between them. If one of these demes were, by chance, to cross over an adaptive valley, he thought it could then transform the rest of the species to the new peak, because it would send out more migrants to 'infect' the other demes. Thus Wright's model (in its later form) is, in effect, a model

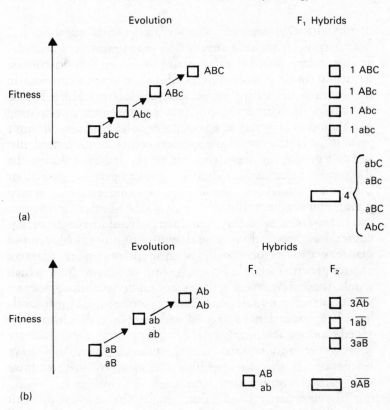

Figure 16.1. (a) *Haploid case*: Three loci. Suppose it increases fitness to be *A*, and to be *B* only if already *A*, and to be *C* only if already *A* and *B*. (b) *Diploid case*: Suppose *A* is dominant to *a*, and *B* to *b*, and that the fitness of the four phenotypes are in the order $\bar{A}\bar{b} > \bar{a}\bar{b} > \bar{a}\bar{B} \gg \bar{A}\bar{B}$.

of phyletic evolution of a whole species that, because of its demic structure, is able to cross adaptive valleys that would be impossible to a large panmictic population. The virtue of the idea is that the crossing of a valley, itself a very improbable event, would be made more likely because of the very large number of demes, any one of which might achieve the crossing. Its weakness is that most species seem not to be divided into demes that are sufficiently isolated for his mechanism to work. However, whether or not one thinks his mechanism plausible, it is, in any case, a mechanism of phyletic evolution of species of large total number, and hence would lend little support to punctuationist ideas.

The other possible role that has been proposed for events in small populations is Mayr's proposal of 'genetic revolutions' occurring in peripheral isolates. Here I face difficulties of interpretation. Thus, suppose that a peripheral population is exposed to environmental conditions different from those of the rest of the species, and that it is sufficiently isolated genetically from the rest of the species to adapt to the new conditions, without being swamped by gene flow from the centre. If isolation lasts for long enough, such a population might well evolve into a new species. However, the characteristics of this new species would have arisen by natural selection; they would evolve more readily if the isolate was not too small, because there would then be greater genetic variability for selection to act on. In no sense would these new characteristics be 'random' with respect to selection within populations. This type of process must often have happened; I have argued earlier that it should not be called 'species selection'.

However, an alternative interpretation of Mayr's 'genetic revolution' is that the peripheral isolate, by virtue of its small size, acquires new characteristics by genetic drift, initially unrelated to selection. If, at the same time, the isolate acquires reproductive isolation from the parent species, we have, indeed, a process of species origin that meets the criteria needed for 'species selection' to be a meaningful concept. A new species has arisen that, simultaneously, acquires reproductive isolation and new characteristics that are not the result of within-population selection.

I doubt very much whether Mayr intended his 'genetic revolution' to be interpreted in this way. Thus, writing of Goldschmidt's concept of 'systemic mutations', according to which a complete genetic reconstruction gives rise in a single step to a well-adapted individual, Mayr accepts the criticism that this is 'equivalent to a belief in miracles'. It would be equally miraculous if a series of changes occurring by chance in a small population were to give rise to a new complex adaptation. In fact, it seems clear that when writing of genetic revolutions, Mayr had in mind the possibility that selection would produce different results in a small population, closed to migrants, than it would in a large one. I think he may have exaggerated the difference, but I doubt if he envisaged the origin of new species as being a sudden event, unrelated to the action of natural selection in populations.

17

Did Darwin get it Right?

I think I can see what is breaking down in evolutionary theory – the strict construction of the modern synthesis with its belief in pervasive adaptation, gradualism and extrapolation by smooth continuity from causes of change in local populations to major trends and transitions in the history of life.

A new and general evolutionary theory will embody this notion of hierarchy and stress a variety of themes either ignored or explicitly rejected by the modern synthesis.

These quotations come from a recent paper in *Palaeobiology* by Stephen Jay Gould. What is the new theory? Is it indeed likely to replace the currently orthodox 'neo-Darwinian' view? Proponents of the new view make a minimum and a maximum claim. The minimum claim is an empirical one concerning the nature of the fossil record. It is that species, once they come into existence, persist with little or no change, often for millions of years ('stasis'), and that evolutionary change is concentrated into relatively brief periods ('punctuation'), these punctuational changes occurring at the moment when a single species splits into two. The maximal claim is a deduction from this, together with arguments drawn from the study of development: it is that evolutionary change, when it does occur, is not caused by natural selection operating on the genetic differences between members of populations, as Darwin argued and as most contemporary evolutionists would agree, but by some other process. I shall discuss these claims in turn; as will be apparent, it would be possible to accept the first without being driven to accept the second.

The claim of stasis and punctuation will ultimately be settled by a study of the fossil record. I am not a palaeontologist, and it might therefore be wiser if I were to say merely that some palaeontologists assert that it is true, and others are vehemently denying it. There is something, however, that an outsider can say. It is that the matter can be settled only by a statistical analysis of measurements of fossil populations from different levels in the rocks, and not by an analysis of the lengths of time for which particular named species or genera persist in the fossil record. The trouble with the latter method is that one does not know whether one is studying the rates of evolution of real organisms, or merely the habits of the taxonomists who gave the names to the fossils. Suppose that in some lineage evolutionary change took place at a more or less steady rate, to such an extent that the earliest and latest forms are sufficiently different to warrant their being placed in different species. If there is at some point a gap in the record, because suitable deposits were not being laid down or have since been eroded, then there will be a gap in the sequence of forms, and taxonomists will give fossils before the gap one name and after it another. It follows that an analysis of named forms tells us little: measurements of populations, on the other hand, would reveal whether change was or was not occurring before and after the gap.

My reason for making this rather obvious point is that the only extended presentation of the punctuationist view – Stanley's book, *Macroevolution* – rests almost entirely on an analysis of the durations of named species and genera. When he does present population measurements, they tend to support the view that changes are gradual rather than sudden. I think that at least some of the changes he presents as examples of sudden change will turn out on analysis to point the other way. I was unable to find any evidence in the book which supported, let alone established, the punctuationist view.

Of course, that is not to say that the punctuationist view is not correct. One study, based on a proper statistical analysis, which does support the minimal claim, but not the maximal one, is Williamson's study of the fresh-water molluscs (snails and bivalves) of the Lake Turkana region of Africa over the last five million years. Of the 21 species studied, most showed no substantial evolutionary change during the whole period: 'stasis' was a reality. The remaining six species were more interesting. They also showed little change for most of the period. There was, however, a time when the water table fell and the lake was isolated from the rest of the rift valley. When this occurred, these six species changed rather rapidly. Through a depth of deposit of about one metre, corresponding roughly to 50,000 years, successive populations show changes of shape great enough to justify placing the later forms in different species. Later, when the lake was again connected to the rest of the rift valley, these new forms disappear suddenly, and are replaced by the original forms, which presumably re-entered the lake from outside, where they had persisted unchanged.

This is a clear example of stasis and punctuation. However, it offers no support for the view that changes, when they do occur, are not the result of selection acting within populations. Williamson does have intermediate populations, so we know that the change did not depend on the occurrence of a 'hopeful monster' (see below), or on the existence of an isolated population small enough to permit random changes to outweigh natural selection. The example is also interesting in showing how we may be misled if we study the fossil record only in one place. Suppose that, when the water table rose again, the new form had replaced the original one in the rest of the rift valley, instead of the other way round. Then, if we had examined the fossil record anywhere else but in Lake Turkana, we would have concluded, wrongly, that an effectively instantaneous evolutionary change had occurred.

Williamson's study suggests an easy resolution of the debate. Both sides are right, and the disagreement is purely semantic. A change taking 50,000 years is sudden to a palaeontologist but gradual to a population geneticist. My own guess is that there is not much more to the argument than that. However, the debate shows no signs of going away.

One question that arises is how far the new ideas are actually new. Much less so, I think, than their proponents would have us believe. They speak and write as if the orthodox view is that evolution occurs at a rate which is not only 'gradual' but uniform. Yet George Gaylord Simpson, one of the main architects of the 'modern synthesis' now under attack, wrote a book, *Tempo and Mode in Evolution*, devoted to emphasising the great variability of evolutionary rates. It has never been part of the modern synthesis that evolutionary rates are uniform.

Yet there is a real point at issue. If it turns out to be the case that all, or most, evolutionary change is concentrated into brief periods, and associated with the splitting of lineages, that would require some serious rethinking. Oddly enough, it is not so much the sudden changes which would raise difficulties, but the intervening stasis. Why should a species remain unchanged for millions of years? The explanation favoured by most punctuationists is that there are 'developmental constraints' which must be overcome before a species can change. The suggestion is that the members of a given species share a developmental pathway which can be modified so as to produce some kinds of change in adult structure rather easily, and other kinds of change only with great difficulty, or not at all. I do not doubt that this is true: indeed, in my book *The Theory of Evolution*, published in 1958 and intended as a popular account of the modern synthesis, I spent some time emphasising that 'the pattern of development of a given species is such that there are only a limited number of ways in which it can be altered without causing complete breakdown.' Neo-Darwinists have never supposed that genetic mutation is equally likely to produce changes in adult structure in any direction: all that is assumed is that mutations do not, as a general rule, adapt organisms to withstand the agents which caused them. What is at issue, then, is not whether there are developmental constraints, because clearly there are, but whether such constraints can account for stasis in evolution.

I find it hard to accept such an explanation for stasis, for two reasons. The first is that artificial selection can and does produce dramatic morphological change: one has only to look at the breeds of dogs to appreciate that. The second is that species are not uniform in space. Most species with a wide geographical range show differences between regions. Often these differences are so great that one does not know whether the extreme forms would behave as a single species if they met. Occasionally we know that they would not. This requires that a ring of forms should arise, with the terminal links overlapping. The Herring Gull and Lesser Black-Backed Gull afford a familiar example. In Britain and Scandinavia they behave as distinct species, without hybridising, but they are linked by a series of forms encircling the Arctic.

Stasis in time is, therefore, a puzzle, since it seems not to occur in space. The simplest explanation is that species remain constant in time if their environments remain constant. It is also worth remembering that the hard parts of marine invertebrates, on which most arguments for stasis are based, tell us relatively little about the animals within. There are on our beaches two species of periwinkle whose shells are indistinguishable, but which do not interbreed and of which one lays eggs and the other bears live young.

The question of stasis and punctuation will be settled by a statistical analysis of the fossil record, but what of the wider issues? Is mutation plus natural selection within populations sufficient to explain evolution on a large scale, or must new mechanisms be proposed?

It is helpful to start by asking why Darwin himself was a believer in gradual change. The reason lies, I believe, in the nature of the problem he was trying to solve. For Darwin, the outstanding characteristic of living organisms which called for an explanation was the detailed way in which they are adapted to their forms of life. He knew that 'sports' – structural novelties of large extent – did arise from time to time, but felt that fine adaptation could not be explained by large changes of this kind: it would be like trying to perform a surgical operation with a mechanically-controlled scalpel which could only be moved a foot at a time. Gruber has suggested that Darwin's equating of gradual with natural and of sudden with supernatural was a permanent feature of this thinking, which predated his evolutionary views and his loss of religious faith. It may have originated with Archbishop Sumner's argument (on which Darwin made notes when a student at Cambridge) that Christ must have been a divine rather than a human teacher because of the suddeness with which his teachings were accepted. Darwin seems to have retained the conviction that sudden changes are supernatural long after he had rejected Sumner's application of the idea.

Whatever the source of Darwin's conviction, I think he was correct both in his emphasis on detailed adaptation as the phenomenon to be explained, and in his conviction that to achieve such adaptation requires large numbers of selective events. It does not, however, follow that all the steps had to be small. I have always had a soft spot for 'hopeful monsters'; new types arising by genetic mutation, strikingly different in some respects from their parents, and taking a first step in the direction of some new adaptation, which could then be perfected by further smaller changes. We know that mutations of large effect occur: our only problem is whether they are ever incorporated during evolution, or are always eliminated by selection. I see no *a priori* reason why such large steps should not occasionally happen in evolution. What genetic evidence we have points the other way, however. On the relatively few occasions when related species differing in some morphological feature have been analysed genetically, it has turned out, as Darwin would have expected had he known of the possibility, that the difference is caused by a number of genes, each of small effect.

As I see it, a hopeful monster would still stand or fall by the test of natural selection. There is nothing here to call for radical rethinking. Perhaps the greatest weakness of the punctuationists is their failure to suggest a plausible alternative mechanism. The nearest they have come is the hypothesis of 'species selection'. The idea is that when a new species arises, it differs from its ancestral species in ways which are random relative to any longterm evolutionary trends. Species will differ, however, in their likelihood of going extinct, and of splitting again to form new species. Thus selection will operate between species, favouring those characteristics which make extinction unlikely and splitting likely. In 'species selection', as compared to classical individual selection, the species replaces the individual organism, extinction replaces death, the splitting of species into two replaces birth, and mutation is replaced by punctuational changes at the time of splitting.

Some such process must take place. I have argued elsewhere that it may have been a relevant force in maintaining sexual reproduction in higher animals. It is, however, a weak force compared to typical Darwinian between-individual selection, basically because the origin and extinction of species are rare events compared to the birth and death of individuals. Some critics of Darwinism have argued that the perfection of adaptation is too great to be accounted for by the selection of random mutations. I think, on quantitative grounds, that they are mistaken. If, however, they were to use the same argument to refute species selection as the major cause of evolutionary trends, they might well be right. For punctuationists, one way out of the difficulty would be to argue that adaptation is in fact less precise than biologists have supposed. Gould has recently tried this road. As it happens, I think he is right to complain of some of the more fanciful adaptive explanations that have been offered, but I also think that he will find that the residue of genuine adaptive fit between structure and function is orders of magnitude too great to be explained by species selection.

One other extension of the punctuationist argument is worth discussing. As explained above, stasis has been explained by developmental constraints. This amounts to saying that the developmental processes are such that only certain kinds of animal are possible and viable. The extension is to apply the same idea to explain the existence of the major patterns of organisation, or 'bauplans', observable in the natural world. The existence of such bauplans is not at issue. For example, all vertebrates, whether swimming, flying, creeping or burrowing, have the same basic pattern of an internal jointed backbone with a hollow nerve cord above it and segmented body muscles either side of it, and the vast majority have two pairs of fins, or of legs which are derived from fins (although a few have lost one or both pairs of appendages). Why should this be so?

Darwin's opinion is worth quoting. In The Origin of Species, he wrote:

> It is generally acknowledged that all organic beings have been formed on two laws – Unity of Type, and the Conditions of Existence. By unity of type is meant that fundamental agreement in structure which we see in organic beings of the same class, and which is quite independent of their habits of life. On my theory, unity of type is explained by unity of descent. The expression of conditions of existence, so often insisted on by the illustrious Cuvier, is fully embraced by the principle of natural selection. For natural selection acts by either now adapting the varying parts of each being to its organic and inorganic conditions of life; or by having adapted them during the long-past periods of time. . .Hence, in fact, the law of Conditions of Existence is the higher law; as it includes, through the inheritance of former adaptations, that of Unity of Type.

That is, we have two pairs of limbs because our remote ancestors had two pairs of fins, and they had two pairs of fins because that is an efficient number for a swimming animal to have.

I full share Darwin's opinion. The basic vertebrate pattern arose in the first place as an adaptation for sinusoidal swimming. Early fish have two pairs of fins for the same reason that most early aeroplanes had wings and tailplane: two pairs of fins is the smallest number that can produce an upward or downward force through any point in the body. In the same vein, insects (which are descended from animals with many legs) have six legs because that is the smallest number which permits an insect to take half his legs off the ground and not fall over.

The alternative view would be that there are (as yet unknown) laws of form or development which permit only certain kinds of organisms to exist – for example, organisms with internal skeletons, dorsal nerve cords and four legs, or with external skeletons, ventral nerve cords and six legs – and which forbid all others, in the same way that the laws of physics permits only elliptical planetary orbits, or the laws of chemistry permit only certain compounds. This view is a manifestation of the 'physics envy' which still infects some biologists. I believe it to be mistaken. In some cases it is demonstrably false. For example, some of the earliest vertebrates had more than two pairs of fins (just as some early aeroplanes had a noseplane as well as a tailplane). Hence there is no general law forbidding such organisms.

What I have said about bauplans does not rule out the possibility that there may be a limited number of kinds of unit developmental process which occur, and which are linked together in various ways to produce adult structures. The discovery of such processes would be of profound importance for biology, and would no doubt influence our views about evolution.

One last word needs to be said about bauplans. They may, as Darwin thought, have arisen in the first place as adaptations to particular ways of life, but, once having arisen, they have proved to be far more conservative in evolution than the way of life which gave them birth. Apparently it has been easier for organisms to adapt to new ways of life by modifying existing structures than by scrapping them and starting afresh. It is for this reason that comparative anatomy is a good guide to relationship.

Punctuationist views will, I believe, prove to be a ripple rather than a revolution in the history of ideas about evolution. Their most positive achievement may be to persuade more people to study populations of fossils with adequate statistical methods. In the meanwhile, those who would like to believe that Darwin is dead, whether because they are creationists, or because they dislike the apparently Thatcherite conclusions which have been drawn from his theory, or find the mathematics of population genetics too hard for them, would be well advised to be cautious: the reports of his death have been exaggerated.

18

Do we need a new
Evolutionary Paradigm?

A claim that a new paradigm is emerging in evolutionary biology cannot be ignored; if it is true, it is important. To establish it, the editors have gathered together a number of essays by authors whose common characteristic is that they do not see themselves as orthodox Darwinists. The reasons for this are various: some have genuine disagreements with neo-Darwinism; some think they do, but only because they have a mistaken view of the Darwinian position; still others are working on subjects to which Darwinian theory is barely relevant. The result, like the proverbial curate's egg, is good in parts, but indigestible if swallowed whole. The contents are so diverse that they can only be reviewed piecemeal.

The first three essays concern the origin of life, and its physical preconditions. Two, by K. Matsuno and by J. S. Wicken, seem to me not so much wrong as vacuous: nothing follows from them other than that life is possible, and we knew that. For some years, people have been measuring energy flows in ecosystems, but, so far, what has emerged is a description, not a causal theory. I think it possible that use of the more sophisticated ideas of non-equilibrium thermodynamics may one day lead to something more interesting: Wicken's essay persuades me that the day is not yet come.

The third essay, by Sidney Fox, is very different. It describes his own fascinating work on some of the lifelike processes that occur in the absence of living organisms, given appropriate conditions. His work may be relevant to a non-Darwinian symposium because it shows that interesting things happen (for example, the polymerisation of amino acids to form proteinoids: non-specific catalysis; the aggregation of proteinoids to form microspheres), without the need for hereditary replication and the resulting evolution by natural selection. I would place greater emphasis than Fox on the origin of hereditary replication as the essential precondition for life, but I recognise this as a field where a less strictly Darwinian approach has had a fruitful outcome.

There follow two essays on 'Pattern and Process'. The first, by Elizabeth Vrba, discusses the different ways in which evolutionary patterns can be explained. For example, suppose that some trait, X, becomes more prevalent. Is this because individuals with trait X outcompete those who lack it? Or because X confers on a species some property which makes it more likely to split into two, or less likely to go extinct ('species selection')? Or because processes not directly affected by X have, as their consequence, an increase in X (the 'effect hypothesis' – for example, if mammals survived the mass extinctions at the end of the Cretaceous because they were homeotherms, a consequence would be the spread of the characteristic mammalian jaw articulation)? Vrba illustrates her argument by examples from the African antelopes: I was interested that she finds no evidence for species selection. Her article differs from most in this volume by being at the same time rational and about evolution. I do not always agree with her, but I would enjoy discussing our disagreements.

The second of this pair of essays, by G. J. Nelson and N. Platnick, discusses the relevance of cladistic taxonomy to Darwinism. Its weakness is that there is no such relevance. Thus the main thrust of their argument is that there is no such thing as an ancestral taxon. If you accept their rule that every taxon (that is, every family, order, class, etc.) must be monophyletic, in the sense of including all the descendants of some ancestor, this is obviously true. For example, the class Reptilia is ruled out, because it does not include the mammals and birds, which are as much descendants of the first reptiles as are tortoises or crocodiles. This is self-evident, but has no bearing on Darwinism, one way or the other. It is an argument about what we should call things, and not about what the world is or was like.

There follow four essays, on development, generative processes and morphogenetic fields. I can discuss only one, and choose P. T. Saunders', mainly because he is an editor. His problem is the development of form. He points out that, in 1952, Alan Turing provided a solution of the problem of how an initially homogeneous field can develop a spatial pattern (typically of standing or travelling waves), arising only from diffusion and chemical reaction, and he argues that such 'prepatterns' are important in development. He argues that the kinds of variation found in any group of organisms will depend on the nature of the developmental processes in that group. Finally, he argues that, if we are going to understand development, we must study these large-scale field processes, as well as the more familiar molecular ones.

He explains these ideas well, and I think he is right. But the essay has some weaknesses. First, there are strange gaps in the references. The first person to emphasise the restricted range of variation in a given taxon was Vavilov, the Russian geneticist who died in a prison he had been sent to for opposing Lysenko. The concept of a prepattern was introduced by the American geneticist Curt Stern. And I was the first person to apply Turing's ideas to evolution, in theory and experiment. We were all hard-nosed neo-Darwinists, and none of us gets a reference. To complain about lack of reference is usually petty, but on this occasion the complaint must be made, because only by omitting us is it possible to suggest that these ideas are 'beyond neo-Darwinism'.

The second weakness concerns Saunders' ignorance of biology. He argues that mimicry illustrates the fact that organisms with similar developmental pathways can rather

easily throw up the same variants. It doesn't. Even in the
same genus, two butterflies may produce the same colour
by means of different pigments. With less closely related
species, it is common to find that a pattern on the body of
the model is copied by features of the wing base of the
mimic. Mimicry between members of different taxa is not
uncommon. Spiders mimic ants, but not because they have
similar developmental processes. Most decisive of all, orchids,
with their fundamentally triradiate symmetry, mimic bilater-
ally symetrical bees. The message of mimicry is that
development does not place any significant constraints on
what can mimic what.

Next are two essays, by Mae-Wan Ho and by J. W.
Pollard, on genetics and evolution. Pollard argues that recent
discoveries in molecular genetics require a radical revision
of Darwinism. In particular, he stresses the work of E. J.
Steele, who proposed a molecular mechanism whereby traits
acquired by individuals during their lives can affect their
offspring – the so-called 'inheritance of acquired characters'.
He proposed the immune system as the obvious first place
to look for such an effect, and later, in collaboration with
G. M. Gorczynski, claimed to have found it. There are at
present two snags. The first, which Pollard discusses, is that
other investigators have failed to observe the effect that
Steele reports, and most immunologists doubt whether any
such process occurs. The second is that, if Steele's process
does operate, it will favour the spread of traits that make
individual cells successful in the struggle with other cells,
but not of adaptations at the level of the organism. The end
result of the process is that we would all be born with
cancer. However, I hope that evolutionary biologists will
read Pollard's essay with care and sympathy. The genome
is a lot more labile than we used to think, and this is not
the time to be dogmatic about what is and is not possible.

The last two essays, by Margaret Boden and by Chris
Sinha, concern artificial intelligence and developmental
psychology. I found them informative and stimulating, but
I am somewhat at a loss to know what they are doing in
this volume. I take it that they are here because they favour
a non-reductionist research strategy. Fair enough, but if that

is sufficient I could suggest other non-reductionists whose work is more relevant to evolution – Ernst Mayr for one.

This volume fails to convince me that a new evolutionary paradigm is needed, let alone that it has arrived. The claim made in the introduction is that we need an organism-based biology. One of the reasons Ho and Saunders offer for this is that M. Kimura's neutral mutation theory has undermined Darwinism. This is an astonishing remark, for two reasons. First, Kimura is not an anti-Darwinist; he says in his recent book: 'the basic mechanism for adaptive evolution is natural selection acting on variations produced by changes in chromosomes and genes.' One cannot get more neo-Darwinist than that. Secondly, in so far as he has departed from Darwinism, he has done so by abandoning the organism, not by espousing it. His theory is that we can understand much of molecular evolution by assuming that changes in molecules have no effect on the organism. The oddest thing about the present book, claiming as it does to usher in a new organism-based biology, is the almost total lack of reference to living animals and plants (I except Vrba; in this respect, as in others, she is a good woman fallen among thieves). It is organism-free biology.

Of course we need to see further than Darwin, but we shall do so by standing on his shoulders, not by turning our backs on him.

PART 4

GAMES, SEX AND EVOLUTION

The essays in this section are close to my own research
interests. Most of my work for the past twenty years has
been concerned with two topics: the evolution of sex, and
the development of evolutionary game theory as a tool for
thinking about evolution. *Why Sex?* was specially written
for this volume: it is the first time I have tried to summarise
my ideas on the subject for a general readership. *The
Limitations of Evolution Theory* was written ten years ago.
I wish I could say that the problems I then listed as unsolved
had received answers, but I fear I cannot. The three reviews,
of books by Dawkins, Gould and Charnov, need no
explanation. *Evolution and the Theory of Games* was an
early attempt to explain these ideas. It has been partly
rewritten, mainly in order to replace some algebra by more
comprehensible arithmetic. It is the only essay in this book
that I have altered, except to remove references and a few
footnotes. Finally, *The Evolution of Animal Intelligence*
attempts to apply game theory to the origins of cooperative
and intelligent behaviour.

19

Why Sex?

We are so used to associating the ideas of sex and reproduction that it is easy to forget that, at a deep level, these two processes are precise opposites. Reproduction is the process in which one cell turns into two, and sex that in which two cells fuse to form one. Darwin has taught us to expect organisms to have properties that ensure successful survival and reproduction. Why, then, should they bother with sex, which interrupts reproduction?

I have spent much of the past twenty years thinking about this problem, which I regard as the most interesting in current evolutionary biology. Yet this is the first time I have attempted to write about it for a general audience. The reason is simple; I am not sure I know the answer. It is always difficult to explain scientific ideas to non-specialists, but it is doubly so if one is not clear about the ideas oneself. So you may find this essay confusing. However, if you are not a biologist you will probably learn some curious and interesting facts.

I must start by explaining why the problem is so difficult. It is not merely that sex seems pointless: it is actually costly. To see why, think of an animal in which the young are not cared for by their parents – to fix ideas, let it be a herring.

Suppose that there are two kinds of female: ordinary sexual females, laying eggs half of which develop into sons and half to daughters, and 'parthenogenetic' females which lay eggs that develop, without fertilisation, into parthenogenetic females like their mothers. If all else were equal, it is clear that the proportion of parthenogens would double in each generation. The problem of sex consists in explaining why all else is not equal. We are seeking a twofold advantage of sex, to counterbalance the twofold cost of producing sons.

In seeking such an advantage, it seems sensible to start with the essential consequence of the sexual process: this is that genetic material from two different individuals is brought together in a single descendant. There is one obvious reason why this might be advantageous. Genetic material gets damaged, by ionising radiation, errors in copying, and so on. In different individuals, different genes will be damaged. If two individuals fuse, each may be able to complement the weaknesses of the other. I have called this the 'engine and gear box' model: one can make one good motorcar by combining the engine from one crock and the gear box from another. It is this process which explains the hybrid vigour that actually results when two inbred strains are crossed. It is also the likely explanation of a process that commonly occurs in fungi, in which two genetically different threads fuse to form a single thread with two different kinds of nuclei. It seems very likely that a short-term advantage, analogous to hybrid vigour, was the evolutionary origin of the cellular fusion that occurs at fertilisation.

If cellular fusion is accompanied by nuclear fusion, as it is in fertilisation, the immediate result is a doubling of the chromosome number, to form a 'diploid'. Clearly, this cannot go on repeatedly, without some alternative process that reduces the chromosome number again. This is what happens in the process of 'meiosis', whereby eggs and sperm are produced. It is the origin and maintenance of meiosis that is our real problem. It is not enough to say that we have to have meiosis because, without it, we could not continue to have fertilisation and the advantages of hybrid vigour. This would be to ascribe foresight to evolution. It amounts to saying that a recently formed diploid undergoes meiosis because, if it did not, it would, over many thousands of generations, accumulate deleterious mutations and therefore need to undergo fusion, but would be unable to do so without a further doubling of the chromosome number. This is the kind of explanation that evolutionary biologists are very reluctant to accept. We expect organisms to have features that ensure their survival and reproduction, and therefore to have the appearance of having been designed, because the process of natural selection will produce precisely such an appearance. But natural selection will favour traits ensuring immediate survival: it cannot favour a trait which will not be advantageous for a thousand generations. I shall return to this problem of foresight later.

For the moment, then, our problem stands as follows. We can understand cellular fusion: it confers the advantage of hybrid vigour. It is harder to understand the process of meiosis, whereby the chromosome number is halved. Let us, therefore, consider another possible advantage of sex. There are good reasons to think that a sexual population can evolve more rapidly to meet changing circumstances. The argument goes as follows. A favourable mutation is a rare event: it requires a specific error in DNA replication, and errors are very rare (each time a base in the DNA is replicated, the chance of mis-pairing is only about one in a thousand million). Suppose two different favourable mutations, A and B, occur in different individuals in a population. Each is likely to increase in frequency, but how can they be brought together in a single individual? Without sex, it is impossible: an AB individual can arise only if a B mutation occurs in an individual that is already A, or vice versa. But A and B could be brought together in the sexual process. The result is that a sexual population can accumulate favourable mutations more rapidly than an asexual one.

According to this argument, then, sex exists because sexual populations evolve more rapidly than asexual ones. There are two objections to the explanation. The first, which I think can be met, is as follows. Is it really necessary for species to evolve rapidly? Most of the time conditions do not change all that quickly, and when they do, as at the onset of an ice age, species tend to meet the change, not by staying put and evolving to meet it, but by moving north or south, so that conditions do not change too much. This may be partly true for changes in physical conditions, but the most important component of the 'environment' for any species consists of all the other species – its competitors, predators and parasites. Each species must evolve, because others are evolving. This has been called the Red Queen model of evolution. As the Red Queen said to Alice, 'It takes all the running you can do, to keep in the same place.'

It seems reasonable, then, that there should be a continuing need for evolutionary change. There is, however, a second objection that is harder to meet. Are we not again ascribing foresight to evolution? We are saying that a species will not abandon sex today, because the environment may change tomorrow. The difficulty can be put in a different way, as a conflict between 'individual selection' and 'species selection'. To revert to our example of the parthenogenetic herring, the case seems to be as follows. In the short run, parthogenetic females would replace sexual ones, because selection would favour them as individuals. Herrings would become an asexual species. Consequently, they would gradually fall behind in the evolutionary race with other species. In the long run, herring would go extinct. Selection would favour sexual species at the expense of asexual ones.

I think this explanation is part of the truth. One reason for thinking so is as follows. There are many parthenogenetic species of animals and plants, but, almost always, these species have close sexual relatives. For example, most Dandelions are parthenogenetic, but there are also sexual species of Dandelions. There are a few exceptions to this rule: the most striking is an order of rotifers (microscopic fresh-water 'wheel animalcules') divided into three families in which no one has ever seen a male. But this is very unusual. What this means is that parthenogenetic species, when they do arise, do not last very long. This is exactly what we would expect on the 'species selection' explanation.

It is clear that species selection can give an appearance of foresight to evolution. It can explain the existence of traits that are disadvantageous to the individual, but advantageous, in the long term, to the species. Such traits will appear to display foresight. However, such an explanation can work only in very special circumstances. What these circumstances are can be seen if we return to the imaginary example of the parthenogenetic herring. I argued that, if a parthenogenetic variety of herring arose, it would rapidly displace the sexual herring, but that the asexual herring would then be replaced by more rapidly evolving sexual species. Therefore, most surviving species are sexual. For this to be true, the origin of new parthenogenetic varieties must be a rare event. If it was a common event, occurring in most species, then most species would become asexual, and there would be a few sexual species left to outcompete them and drive them to extinction. A species-selection explanation will work only if new parthenogenetic varieties arise very rarely.

For parthenogenesis, this is probably the case. The basic reason is that, starting from sexual ancestors, the origin of parthenogenesis requires several simultaneous changes, and not just one. In many sexual species, a female lays few eggs until she has been mated. Eggs usually do not start developing unless they are stimulated to do so by sperm entry. Successful development usually requires that the egg be diploid: this can be achieved either by suppressing meiosis, or allowing meiosis and then doubling the chromosome number again. Hence the origin of a successful parthenogen requires that female egg-laying behaviour should change, that eggs should start development without being stimulated by sperm, and that diploidy be maintained. For these and other reasons, new parthenogenetic varieties do not occur often.

Thus a new parthenogenetic variety must overcome many hang-ups, that exist only because of its recent descent from a sexual ancestor. Some idea of these sexual hang-ups can be obtained by reviewing the occurrence of parthenogenesis among vertebrates. Among the lower vertebrates, a number of fishes, frogs and salamanders are parthenogenetic. In no case, however, is there a parthenogenetic species capable of independent existence, because the eggs laid by the parthenogenetic female always require the stimulus of sperm penetration to start development: the sperm are obtained by mating with males of a related sexual species. Typically, the sperm contributes no chromosomes to the new individual.

Thus these strains are true parthenogens, but they will never replace their sexual relatives unless they can dispense with their reliance on sperm to trigger development. The sexual males, of course, are wasting their time, but have not yet evolved complete avoidance of the parthenogenetic females.

A still more bizarre situation occurs in a few vertebrates, of which the green marsh frogs are an example. *Rana esculenta* is a hybrid between two sexual species, *R. ridibunda* and *R. lessoni*. However, the eggs it produces contain only the chromosomes from *R. ridibunda*: the *R. lessoni* chromosomes are eliminated. The hybrids mate with males of *R. lessoni*, which provide a new set of *lessoni* chromosomes, destined also to be discarded when the next generation of eggs are provided. Again, the males are wasting their time, and again, this is probably not a step on the road to an independent parthenogenetic species.

In lizards, sperm seems not to be needed to trigger development. In two genera, *Cnemidophorus* in America and *Lacerta* in Russia, there are fully independent parthenogenetic species consisting entirely of females that produce offspring genetically identical to themselves. The 'species' *Cnemidophorous uniparens* apparently consists of a single clone: scales can be successfully grafted between any two individuals. This suggests that the species is descended from a single female, probably a hybrid between two sexual species, and that it is relatively recent – perhaps a few thousand years.

In birds and mammals, no wild parthenogens are known. There are very sickly domestic strains of turkeys and chickens that are parthenogenetic, but we are a long way from a successful all-female strain of egg-laying chickens, which is what the breeders would like. In mammals, there is no reliable report of parthenogenesis, even with modern watch-glass technology. In a recent article in *Nature*, I said that I knew of no explanation for this absence, but one possibility has since been pointed out to me. It seems that in mammals the genes inherited from the father and the mother are, in some way, differently labelled or imprinted, and play different roles in early development: this, of course, runs counter to the typical Mendelian scheme. In consequence, a mammalian egg containing two haploid nuclei will develop only if one nucleus is derived from a male and one from a female. If this is correct, it accounts for the absence of parthenogenesis in mammals.

From the brief review of parthenogenesis in vertebrates, it seems that only in a few lizards are there parthenogens that stand any real chance of selectively displacing their sexual ancestors. This confirms that the origin of effective parthenogenesis is a rare event. In fact, it may provide misleadingly strong evidence for this view. In higher plants, parthenogenesis is by no means unusual – Blackberries, Dandelions and Hawkweeds are familiar examples. In arthropods too, parthenogenesis is quite common. The most familiar are the Aphids (greenfly and blackfly) and Cladocerans (water fleas). In both these groups, parthenogenesis is cyclical: asexual reproduction alternates with the production of males and females, and a sexual generation. However, in both groups, a number of strains have wholly abandoned sex.

Groups such as the Dandelions, Aphids and Cladocerans do suggest that something more is needed than the long-term species selection argument. The origin of new parthenogenetic clones seems to be too common an event to be kept in check by species extinction. This has led to the idea that there is some short-term advantage to sex, arising from the genetic diversity it generates, and the advantage that such diversity confers in an environment that varies from place to place. There may be something in this idea, but it is not as obvious as it might seem. G. C. Williams expressed it with a vivid analogy. An asexual parent, he said, is like a man who buys 100 tickets in a raffle, and finds that they all carry the same number. The point he is making, of course, is that the offspring of a parthenogenetic female are (in most cases) genetically identical to one another. But suppose the 100 offspring of a parthenogen are widely distributed in space. Then the parthenogen is like a man who buys 100 tickets, all with the same number, but in 100 different raffles. There is no harm in that.

It may be that I have been approaching the problem from entirely the wrong direction. I have been asking why sex is retained, despite its twofold disadvantage, and attempting to answer the question by looking at higher animals and plants. Perhaps I should have been asking why sex arose in the first place, and have been looking at simple single-celled organisms. The major subdivision that is today recognised among living organisms is that between 'prokaryotes' and 'eukaryotes'. The former, which include the bacteria and blue-green algae, have no cell nucleus, and a single ring-shaped chromosome: the eukaryotes include all the rest, including ourselves, and oak trees, and many single-celled organisms, like amoeba and the parasites that cause malaria and sleeping sickness.

It seems a reasonable guess that our ancestors were once at a prokaryote stage of complexity. What can we learn from them about our own sexual habits? I suspect, rather little. There is no process of meiosis or fertilisation in prokaryotes. There are a number of ways in which segments of DNA pass from one cell to another. However, I think these DNA segments are better thought of as parasites or symbionts than as analogues of sex. One example is worth describing in a little more detail. There is a DNA element known as an F plasmid. A bacterial cell containing an F plasmid develops hair-like 'pili' on its surface, which attach it to other bacteria. When this happens, a copy of the F plasmid is transferred to the second bacterium. This looks like a simple case of a parasite arranging for its own transmission – just as a flu virus makes you sneeze. However, in a small proportion of cases, a portion of the bacterial chromosome may be transferred instead. This does look more like a sexual process, but I do not think it is ancestral to eukaryotic sex. As far as sex is concerned, the main thing we inherited from our prokaryotic ancestors was a battery of enzymes which originated, and which still serve, to repair DNA damaged by ionising radiation, but which also serve to recombine chromosomes during meiosis.

The eukaryotes are thought to have originated by the symbiotic union – a sort of permanent coalition – of several kinds of prokaryote. Sex as we know it, involving meiosis and fertilisation, is a later development. The reasons for its origin and maintenance are still a matter of controversy. There are two main reasons for the difficulty, one practical and one theoretical. The practical difficulty is that, in all probability, sex originated only once, almost a thousand million years ago. The theoretical difficulty is that we have to think about selection acting on different time scales, and at different levels: that is, between genes, between individuals, and between species.

However, once we accept that sex is typical of the eukaryotes, a number of secondary questions arise, which are a good deal easier to answer. The first concerns the origin of gender – that is, of a differentiation between the two sexes. A female is an organism that produces large non-motile gametes, and a male one that produces small motile ones. This is not the primitive state. In the first sexual organisms, and in many simple organisms today, all gametes are small and motile. As Genesis suggests, males were the first sex. The evolution of a second sex producing large non-motile gametes is a later development. Various intermediate stages can be observed in existing green algae – for example, among the Volvocales, which form spherical green colonies. The reason for the change, and its association with large adult size, is well understood.

Once such a differentiation exists, it is easy to see why there can be only two sexes. Thus suppose there were three sexes, A, B and C. There are two possibilities. First, any two of the sexes can combine to produce a new individual: that is, $A \times B$, or $B \times C$, or $C \times A$. If so, it is inconceivable that the three kinds would be equally efficient: one would be eliminated by selection. Second, all three are needed: that is, $A \times B \times C$. If all three contributed genetic material, this would require a 'meiosis' which reduced the chromosome number to one-third its initial value. Although not inconceivable, this has never happened, and it is hard to see what advantage it would confer. More plausibly, only two of the 'sexes' contribute genetic material, and the third helps to raise the offspring. This has several times been proposed in science fiction. It also happens among animals, if queens, drones and workers in social insects are regarded as different 'sexes'.

An alternative to separate sexes is hermaphroditism, in which both eggs and sperm are produced by the same individual, either simultaneously or in succession. Hermaphroditism is the rule among higher plants, and in some animal groups (for example, snails and flatworms), but is unknown in birds and mammals. There is an interesting economic argument predicting which species should be hermaphrodites. If there is a 'law of diminishing returns' in producing eggs, or producing sperm or pollen, it pays to be hermaphrodite. For example, a plant which produced twice as many seeds would not therefore produce twice as many offspring, if those seeds fell close to the parent plant, and competed with one another for limited space. In contrast, if there is an increasing return on investment in male (or female) function, it pays to be single-sexed. For example, a stag that invested half as much in growth, antlers and fighting ability would get less than half as many matings. In general, the economic argument works rather well. But some higher plants, like holly trees and hops, do have separate sexes: the most likely explanation is that this is a device that prevents self-fertilisation.

I assumed earlier that the numbers of males and females are equal, and this is usually the case. Why should this be so? A short-term answer would be that, in most cases, the machinery of meiosis, combined with a particular genetic sex-determining mechanism, ensures that it is so. But if some other ratio were favoured by selection, I have little doubt that organisms would have found some means of producing it. Why, then, should the 1:1 ratio be favoured? In most species which lack parental care, the productivity of the population would be increased if there were more females than males. The explanation is that selection acts at the individual level. Thus suppose that females could choose the sex of their children, and that their choice should be that which maximises the number of grandchildren they have. Which sex should they choose? If most other females are producing sons, they should produce daughters, and vice versa. This follows from the fact that each child has one father and one mother, and hence that, on average, members of the rarer sex have more children. The only stable state arises when the numbers of males and females are equal.

The reason for thinking that the argument in the previous

paragraph is along the right lines is that there are many situations in which a similar line of argument predicts a sex ratio that is not 1:1, and observation shows that the actual sex ratio is distorted in the predicted direction. One last point on the sex ratio. We are accustomed to thinking that sex is determined by whether the sperm carries an X or a Y chromosome, because that is what happens in people. However, it is not universal. In birds and butterflies, sex is determined by the chromosomes in the unfertilised egg, not in the sperm. In tortoises, and some other reptiles, the sex of a new individual is determined by the temperature at which the egg is incubated.

Females and males are defined according to whether they produce eggs or sperm. In some animals there are few other differences between them, but in others there are many. Some differences, such as the presence of mammary glands in female mammals, are directly associated with different roles in reproduction, but many are not. The sexes often differ in size, colour, ornamentation and weapons. Darwin was particularly interested in differences of the last kind, because they are not readily explicable in terms of natural selection. To take an extreme example, why should selection have favoured peacocks with such elaborate and brightly-coloured tails? If such tails really increased fitness, one would expect them to be present in females also. In any case, common sense suggests that a bird with such a tail would be more vulnerable to predators.

It was to account for such traits that Darwin conceived of the process of sexual selection. He thought that this could act in one of two ways – through competition between males for females, and through choice by females of particular kinds of male. Before continuing, it is worth asking why one should see things this way round. Why not competition between females for males, and choice by males of particular females? I can best answer this question by describing the animal on which I worked for some twenty years. This is *Drosophila subobscura*, the commonest European fruit fly. Females mate only once. In a single mating they can acquire enough sperm to fertilise all the eggs (about 1000) they can lay in a lifetime. They are very selective about whom they will mate with, insisting on young, healthy, outbred males and rejecting old or inbred ones. How they do this I shall describe later. In contrast, males will court and attempt to copulate with any object of about the right size (for example, a blob of wax about 2 mm long) that is moved in an appropriate way. Unlike a female, a male can mate about six times a day, and will do so if given the chance. The reason for these differences is obvious. A female's breeding success is limited by the number of eggs she can lay, and would not be increased if she mated repeatedly: a male that mates twice as often will leave twice as many descendants.

This pattern is typical in species in which males do not care for their young. It is to such species that the concepts of male–male competition and female choice best apply. In monogamous species in which both parents care for the young, (in many birds, for example, and in gibbons and marmosets among primates), there are likely to be elements of both competition and choice among both sexes. The opposite extreme from *Drosophila* is reached in the Jacana (a bird that walks on the leaves of water lilies). All the nest-building, incubation and care of the young is done by the males. Therefore a female that can acquire several males will leave more descendants. Accordingly, females are more brightly coloured than the males, and half as heavy again. They defend territories against each other, each territory including the nests of several males.

No great difficulty arises in explaining why male deer and elephant seals, and female Jacanas, compete with members of their own sex to acquire more members of the opposite sex. It is harder to see why female fruit flies, and peahens, should choose to mate with one male rather than another. In *Drosophila subobscura*, we do have some idea of both how and why the choice is made. If a male and a virgin female are placed in a small container, the male will soon notice the female, and approach her head to head. The female then performs a rapid side-to-side dance, and the male also sidesteps so as to keep facing her. If he succeeds, the female stands still and raises her wings, and the male circles and mounts. But an old or inbred male will usually fail to keep up during the dance: if so, the female does not accept him, but walks or flies away. In such a case, one can watch a long succession of dances, none of which end in mating.

It seems, then, that the female is using a test of athletic ability to decide whether to mate with a particular male. Why should she bother? The male is not going to care for the eggs, so why should it matter whether he is physically vigorous? I think that the explanation is as follows. A male that lacks vigour may do so for genetic reasons: for example, he may carry one or more harmful mutations. By avoiding such males, a female makes it more likely that her offspring will survive. This is referred to as a 'good genes' explanation. We do not have much direct evidence that it is correct, but it does make sense.

What about the Peacock's tail? At first sight this is much more puzzling. If, as seems reasonable, the possession of a long tail makes it less likely that a male will survive, surely a female is foolish to prefer such a male? Darwin supposed that females had some kind of aesthetic preference for particular kinds of male, but was unable to offer a convincing explanation of why such preferences should exist. A possible answer was given in 1930 by R. A. Fisher. Suppose that, for whatever reason, the females in some population do in fact prefer males with long tails. Long-tailed males will get more matings, and leave more descendants. What would happen to a female that went against the trend, and mated with a male with a short tail? Her sons would tend to have short tails, and would therefore get few matings. Hence a female that bucked the trend would leave fewer descendants. Therefore, Fisher thought, a kind of runaway process would occur, with females preferring males with longer and longer tails, and males developing such tails. What could get such a process started? Fisher thought that at first the trait chosen by females – for example, a slightly longer tail – would actually increase the male's chance of survival, and that only later would the trait, and the preference, be exaggerated to non-adaptive levels.

This is an ingenious idea. Much more recently, it has been shown mathematically that Fisher's process can work: that is, his assumptions do lead to the results he expected. But that is not the end of the problem. Many biologists are reluctant to accept that the traits chosen by females are in a sense arbitrary, and that the results of female choice are often maladaptive. They have suggested that the chosen trait may be an outward and visible sign of superior fitness. In the case of the dancing ability of male fruit flies, this is easy to accept. Perhaps we can extend the idea to the peacock's tail. Anyone who has watched a peacock raising its tail will know that it is quite a performance: perhaps a weak or unfit male could not manage it so successfully. If so, by choosing a male with a long tail, a female is ensuring that she chooses a male of high vigour. The suggestion is being hotly debated. The two cases are not quite the same. A male fruitfly does not pay any obvious cost in survival by being good at dancing, but a peacock does seem to pay such a cost by growing that tail. What we need now is that more people should study species in which one or other sex has an apparently expensive sexual ornament.

It is hard to study sex in animals without seeing the parallels between animals and people. I find it impossible to talk to non-biologists about these problems without it being assumed that my motive is to understand human sexuality. That may be my motive, but if so it is an unconscious one. I watched the courtship dance of *Drosophila* almost by accident. I was studying the effects of inbreeding, and wanted to find out why so many eggs laid by inbred females failed to hatch. So I watched courtship to make sure that mating had taken place. When (if the male was inbred) it did not, I had to ask why. I stumbled over the phenomenon of female choice: I was not looking for it. My more general interest in the evolution of sex came about because of the apparent involvement of 'species selection'. Sex was, and to some extent still is, something of an anomaly for an orthodox Darwinist, and I think it is a good idea to study anomalies.

That is not to say that I do not recognise the parallels between people and animals. I cannot watch most animal courtship without some element of identification. In fact, I do not even try to do so. To watch animal behaviour without using human analogies would be to rob oneself of a major source of insight. Of course, any insight gained in this way must be tested, and will often prove to be wrong. But more scientists are held up in their research by a shortage of ideas to test than by an inability to test the ideas they have got. Only people who have no intention of studying animal behaviour are likely to reject the human analogy.

But I do see a real danger in arguing in the other direction, from animals to man. This is particularly so if it is argued that what happens in animals is natural, and that what is natural is right. There are two links in that chain of argument, both suspect. The sexual behaviour of animals is so varied that it is never safe to assume that what happens in some animal is natural for people. The behaviour of the Jacana and the Elephant seal are, as far as sex roles are concerned, almost exact opposites: they cannot both be natural for people. In any case, what does 'natural for people' mean? At most it might mean something like 'what most of our ancestors have been doing for the last half million years'. We have, of course, very little idea what they have been doing, although we do have some hints, and the question is well worth asking. Most important of all, even if we did know what they were doing, that would not make it right for us.

20

The Limitations of Evolution Theory

There are a lot of things we do not know about evolution, but they are not the things that non-biologists think we do not know. If I admit to a non-biological colleague that evolution theory is inadequate, he is likely to assume at once that Darwinism is about to be replaced by Lamarckism and natural selection by the inheritance of acquired characters. In fact, nothing seems to me less likely. In common with almost everyone working in the field, I am an unrepentant neo-Darwinist. That is, I think that the origin of evolutionary novelty is a process of gene mutation which is non-adaptive, and that the direction of evolution is largely determined by natural selection. I am enough of a Popperian to know that this is a hypothesis, not a fact, and that observations may one day oblige me to abandon it, but I do not expect to have to. Indeed, everything that has happened during my working life as a biologist, and in particular the development of molecular biology, has strengthened rather than weakened the neo-Darwinist position.

The difficulties of evolution theory seem to me to arise from another direction. The essential components of the theory are mutation (a change in a gene), selection (differential survival or fertility of different types) and migration (movement of individuals from place to place). The theory tells us that each of these processes, at a level far too low to be measurable in most situations, can profoundly affect evolution. For example, consider selection. Suppose that there are two types of individual in a population, say red and blue, which differ by 0.1 per cent, or 1 part in 1000, in their chances of surviving to breed. If the population is reasonably large (in fact, greater than 1000), this difference in chances of survival will determine the direction of evolution, towards red or blue as the case may be. But if we wished to demonstrate a difference in the probability of survival during one generation – that is, we wished to demonstrate natural selection – we would have to follow the fate of over one million individuals, usually an impossible task.

Similar difficulties of measurement arise with mutation and migration. When a gene replicates, there is a chance of the order of 1 in 100 million that a particular base will be miscopied. It is possible to measure these astonishingly low rates of error in very special circumstances in some micro-organisms. It is also clear from the theory that rates of this order are sufficient to provide the raw material of evolution. But in most natural situations, mutation rates cannot be measured. Finally, consider migration. Suppose that a species is subdivided into a number of populations, and we wish to know how far the evolution of any one population is influenced by immigration from the others. Theory shows that if a population receives on the average one migrant from outside in each generation, this can have a decisive effect. Yet in practice we could not hope to measure such a low rate of migration.

Thus we have three processes which we believe to determine the course of evolution, and we have a mathematical theory which tells us that these processes can produce their effects at levels we cannot usually hope to measure directly. It is as if we had a theory of electromagnetism but no means of measuring electric current or magnetic force. Now things are not quite as bad as I have painted them. Surprisingly often, it is possible to measure natural selection directly, because the selective differences are not 1 in 1000, but 1 in

10 or thereabouts. But the measurement difficulty is serious. It means that we can think up a number of possible evolutionary mechanisms, but find it difficult to decide on the relative importance of the different mechanisms we have conceived. I hope that this will become clearer if I give some examples.

1. What determines the rate of evolution?

Suppose that at a time 200 million years ago, during the age of reptiles, some event had occurred which doubled the rate of gene mutation in all existing organisms; we must also suppose that for some reason the rates did not fall back to their original levels. What would have been the consequences? Would the extinction of the dinosaurs, the origin of mammals, of monkeys and of man have taken place sooner, so that roughly the present state was reached in only 100 million years? Or would the rate of evolution have stayed much the same? Might it even have been slower? The short answer is that we do not know.

The difficulty is that we can picture evolution in two ways. According to the first picture, a species spends most of its time more or less 'at rest' in an evolutionary sense; the species is well adapted to the contemporary environment, and selection acts to maintain its characteristics rather than to change them. Occasionally some change in the environment – an ice age or the appearance of a new predator – imposes new selective pressures, and the species responds by a burst of evolution. To do so, it does not have to wait for new mutations, because the species will already have a large amount of genetic variation which was generated by mutation while it was at rest. If this picture were true, an increase in the mutation rate would have little effect on the rate of evolution, which depends on environmental change.

I think this picture arises from our knowledge of the effects of artificial selection on domestic or laboratory animals. If you take a population of, say, mice and select for increased size, the average size will increase rapidly, far more rapidly, by a factor of perhaps 100,000, than metric characters of this kind normally change in evolution. It can be shown that the change does not depend on new mutations occurring since the start of selection, but on genetic differences already present in the population from the outset. What is true of size in mice is true of almost any character you like to name in any outbreeding species. To base a picture of evolution on such experiments, however, overlooks two facts. First, in evolution a number of different characteristics usually change simultaneously, and second, changes in response to artificial selection do not continue indefinitely, but slow down and stop when the initial supply of genetic variability has been used up.

An alternative picture of evolution has been suggested by Van Valen. He points out that the main feature of the 'environment' of most species consists of other species, which it eats, or which eat it or compete with it for food. If so, when any one species in an ecosystem makes an evolutionary advance, this will be experienced by one or many other species as a deterioration of their environment; their food will have got more difficult to find or their predators more difficult to escape. If so, the other species will evolve in their turn, and cause environmental deterioration for still other species. The result will be that every species will be evolving as fast as it can, simply to keep up with the others. Van Valen calls this the Red Queen hypothesis; the Red Queen, you will remember, told Alice: 'Now here, you see, it takes all the running you can do to keep in the same place.'

If this picture is true, then our original question reduces to this; if the mutation rate is doubled, does this double the maximum rate at which a species can evolve? I wish I could tell you that this question has a simple and agreed answer, but I cannot. My own opinion, however, based on some rather messy algebra, is that it does, provided that the population is not very large (roughly, not greater than the reciprocal of the mutation rate; if it were greater than this, then every possible mutation would occur in every generation anyway). But my reason for raising the question in the first place was not to answer it. It was to make the point that a theory of evolution which cannot predict the effect of doubling one of the major parameters of the process leaves something to be desired.

This discussion raises two other questions. What determines the mutation rate itself? Do species go extinct because they cannot evolve fast enough? As I shall show, we do not know the answer to either of them.

2. *What determines the mutation rate*

The mutation rate – that is, the probability of error in gene replication – is not a fundamental constant like Planck's constant or the velocity of light. It depends on the enzymes which replicate DNA and which correct errors as they go. Since the enzymes are produced by genes, the mutation rate is under genetic control.

Two views can be taken about the mutation rate. According to the first, the mutation rate is as low as natural selection can make it; after all, 1 in 100 million per base replication is fairly low. The logic behind this is that the vast majority of mutations are harmful, so an individual with a low mutation rate will leave more offspring.

The alternative view is that the actual mutation rate we observe is an optimum. If it were higher then the harmful effects just mentioned would predominate, but if it were lower there would be some other disadvantage; for example, too low a mutation rate would inhibit evolutionary adaptation.

In some micro-organisms (some bacteria and yeasts) there is now a strong reason for preferring the second view. Genetic strains can be isolated which have a lower mutation rate than typical 'wild' strains. Since a lower mutation rate is possible, the actual rate must be an optimum rather than a minimum. Of course, this argument would fail if the strains with a lower mutation rate suffered some other disadvantage; for example, they might replicate DNA very slowly, just as a neurotically careful proofreader would slow down publication of this article. As far as we can see, there is no corresponding disadvantage in the low mutation-rate strains, but we cannot be sure. More serious, there is no way at present of saying whether what is true of bacteria and yeasts is true of all organisms.

3. Why do species go extinct?

A corollary of the Red Queen hypothesis is that those species which are unable to evolve rapidly to meet changed circumstances will go extinct. But is this an important reason for extinction? I recently wrote a mathematical analysis of the Red Queen hypothesis, and sent a copy of it to Dr G. C. Williams. He replied as follows. Suppose, he said, there had been a flea peculiar to the passenger pigeon. As the pigeons became rarer, the fleas would have been under stronger and stronger selection pressure to exploit an ever-dwindling resource. But no matter how rapidly they improved their adaptation to their way of life, nothing could prevent their extinction.

When a species goes extinct, could it have survived if it had been able to evolve more rapidly? Or does extinction happen because the way of life of a species disappears and there is no other sufficiently similar way of life to which it can adapt? Clearly, both kinds of extinction must sometimes happen. But which is more frequent? We have no idea. As so often in evolution theory, we can imagine two mechanisms but cannot estimate their relative importance. There is, however, a reason why we would like to know more about the causes of extinction; as we shall see in the next section, Williams had a reason for inventing the flea on the passenger pigeon.

4. Sex, altruism and the origins of society

In most discussions of evolution, the 'unit of selection' is the individual organism. We explain the evolution of horses' legs by arguing that those individual horses which had legs of a particular form were more likely to survive and to produce children like themselves. But in some discussions of evolution a different argument is used, or at least implied. For example, the origin and maintenance of sexual reproduction has been explained by saying that species (such as the dandelion, or the lizard *Lacerta saxicola*) which consist entirely of parthenogenetic females cannot evolve so rapidly to meet changed conditions, and so in the long run will go extinct. In this argument, the 'unit of selection' is the species; it is not the individual which evolves or goes extinct, but the species.

Selection depending on the differential survival of species or other populations has been called 'group selection'. Whenever a characteristic is explained as conferring an advantage on a species, rather than on the individuals which compose it, group selection is being assumed. The snag with such explanations is that the process is a very weak one: if there is a conflict between individual and group selection, individual selection is likely to win. This has led people to seek some short-term individual advantage for sexual reproduction. G. C. Williams has been one of the strongest opponents of group selection, which may be why he is reluctant to think that species go extinct because they cannot evolve fast enough.

For the present, it is fair to say that we do not have an agreed explanation of how sex originated or how it is maintained. The same difficulty arises for all aspects of the 'genetic mechanism' – for example, the amount of recombination between genes, the number and arrangement of chromosomes, the pattern of self- and cross-fertilisation, and the mutation rate. In discussing these features, biologists have fallen into the bad habit of assuming that if a particular property can be shown to be advantageous to the species, then this will account for its evolution, without asking themselves whether a species advantage could overcome an individual disadvantage. Discerning readers will have noticed that I used, deliberately, an argument of this kind myself when discussing whether the mutation rate is optimal. Loose thinking of this kind is found not only in discussions of genetic mechanisms, but also of ecology and of animal behaviour.

A related difficulty arises because it is not individuals

which reproduce but genes. The individual is simply a device constructed by the genes to ensure the production of more genes like themselves. Normally, a gene will only increase in frequency if it produces an individual of greater than average fitness; to this extent, the argument above about the legs of horses is sound. But sometimes the multiplication of genes is not tied to the multiplication of the individuals carrying those genes. This is particularly true in bacteria; consider, for example, the 'resistance transfer factors' which can transfer the genes for drug resistance from one species of bacterium to another. The evolutionary processes of bacteria are still a largely unexplored field, but it is one I do not feel competent to discuss.

The fact that it is genes and not individuals which replicate has some intriguing consequences for the evolution of higher organisms. I can best introduce the idea by quoting a remark of J. B. S. Haldane's; he announced that he was prepared to lay down his life for two brothers or eight cousins. The point behind the remark is as follows. Suppose there existed a mutant gene which caused any individual carrying it to be prepared to sacrifice its life if, by so doing, it could ensure the survival of more than two brothers which otherwise would have died. Such a gene would increase in frequency in the population. This is because there is a resemblance, an 'identity by descent', between the genes of relatives; in particular, half my brother's genes are identical by descent to my own.

Thus a gene which reduces the probability of survival of an individual carrying it but produces a corresponding increase in the fertility or probability of survival of relatives can increase in frequency. The process has been called 'kin selection'. It has been used to explain apparently 'altruistic' behaviour (strictly, 'nepotistic' would be a better word), such as the giving of alarm notes by birds or the self-sacrificing behaviour of worker bees. That kin selection will affect evolution would be admitted by anyone who understands the laws of inheritance and who is capable of logical thought. But it is much harder to say whether the process has been an important one. Opinions vary from those who regard it as, at the very most, relevant to

understanding a few peculiar special cases, to those who see
it as a major key to an understanding of animal societies,
including our own.

5. Drift or selection?

I find that I have so far failed to mention the most protracted
debate in evolution theory, which concerns the relative
importance of chance and selection. This is because I find
the argument less interesting than the ones I have discussed.
It does, however, illustrate the point that our difficulty in
evolution theory is not so much to think of processes which
would explain what we see, as to evaluate the relative
importance of different possible processes.

The basic problem is as follows. In an infinite population,
if one can imagine such a thing, natural selection would
always determine which of two types would be established
and which eliminated. But real populations are finite, and
in a finite population it is quite possible for the fitter of two
types to be eliminated and the less fit established, provided
that the fitness differences are small. Such random changes,
not produced by selection and sometimes contrary to
selection, are referred to as 'genetic drift'. There is no
question that drift will occur, but deep disagreement about
its importance. In recent years the argument has taken a
new form, under the titles 'neutral mutation theory' or 'non-
Darwinian evolution'. It has been argued, particularly by
Kimura, that a large part of the variation which we observe,
both within and between species, in the structure of proteins
has no effect on survival, but has arisen by drift.

The question is difficult to answer because selective forces
too small to measure would still be more important than
genetic drift in determining genetic change. It is sometimes
difficult to understand the heat which has been generated,
since both sides, neutralists and selectionists, agree that the
evolution of adaptations is the result of natural selection,
and it is adaptation which is the most striking feature of the
living world.

6. Evolution and development

So far I have discussed difficulties which are in a sense internal to evolution theory. There are other difficulties which arise because of ignorance in other fields of biology, and whose solution will depend on research in those fields. This is most obviously true of the relation between evolution and development.

During the last century ideas about the development of individuals and the evolution of populations were inextricably mixed up in people's minds. When Weismann formulated his theory of the independence of germ line and soma – that is, that the processes whereby a fertilised egg gives rise to germ cells which are the starting point of the next generation is independent of the process whereby a fertilised egg gives rise to an adult organism – he was in effect saying that it is possible to understand genetics without understanding development. He thus set the stage for the growth of the science of genetics during this century; sadly, we still do not understand development. It follows that our understanding of evolution is necessarily partial, because genes are selected through their effects on development.

This difficulty arises in many ways. The most topical concerns the evolution of 'structural' and 'regulator' genes. The term 'structural gene' can be misleading; it does *not* mean a gene concerned with an aspect of adult morphological structure. A structural gene is a gene which codes for a protein, which may be an enzyme or which, like collagen or the muscle proteins, may form the substance of the body. We now know a good deal about the evolution of particular kinds of proteins – for example, the haemoglobins which carry oxygen in our blood, or the hormone insulin – and can therefore infer a good deal about the evolution of structural genes. 'Regulator' genes are a much vaguer concept, particularly in higher organisms. We know that there is a lot more DNA than is needed to code for proteins, and we guess that some of the rest is concerned with regulating the activities of other genes, and so controlling cellular differentiation and growth. In higher organisms we know rather little about how such regulation works, and even less about how it evolves. But it seems clear that structural and regulator genes evolve at very different rates. One rather striking feature of the evolution of structural genes is that the rate of evolution is surprisingly uniform in different groups of organisms, at least for a given class of structural genes – for example, those coding for haemoglobin. The reason for this uniformity of rate is a matter of controversy, but the uniformity itself seems well established. For example, frogs are, as a group, about twice as ancient as placental mammals, and the range of differences between their proteins is correspondingly about twice as great. Yet in morphological structure the differences between mammals are far greater. This presumably reflects a more rapid change in their 'regulator' genes, although we have no direct evidence. Incidentally, it also seems to be true that the numbers and shapes of chromosomes change much more rapidly in mammals than in frogs, but the relation between this and morphological change is still obscure.

These difficulties arise because, although we have a very clear idea about how genes specify proteins, and can therefore speak with confidence about the evolution of proteins, we have no corresponding idea about how genes specify adult morphological structures.

7. How will our ignorance be diminished?

If it were easy to answer the questions I have raised, they would have been answered long ago. It is rarely possible in evolution theory to think of a single decisive experiment or observation which will settle a controversy. Our understanding of evolution depends on a combination of clearly formulated theories and wide comparative knowledge. For example, to understand the evolution of genetic mechanisms, we need on the one hand clear theories which predict that particular types of mechanism should evolve in particular situations, and on the other we need comparative knowledge of the associations between the ecology and taxonomy of particular species and their genetic systems. Unfortunately, the kind of scientist who is good at developing clear theories often finds it difficult to remember facts, whereas those who know the facts tend to jib at the algebra. It seems to me that there is no single idea in biology which is hard to understand, in the way that ideas in physics can be hard. If biology is difficult, it is because of the bewildering number and variety of things one must hold in one's head.

21

The Evolution of Animal Intelligence

The last twenty years has seen a major effort, theoretical and observational, to understand the evolution of animal societies. In this essay, I first discuss what level of intelligence is assumed in our theories, and what level is revealed by our observations. I then ask what qualitative differences exist between animal and human societies, and in what ways these differences depend on human intelligence.

The two leading concepts in the analysis of animal societies are kin selection and evolutionary game theory. The central idea of kin selection is that a gene A, causing an animal to be more likely to perform an act X, may increase in frequency in a population even if act X reduces the individual fitness (expected number of offspring) of the animal itself, provided that the act increases the fitness of animals related to the actor. Following Haldane, biologists refer to such acts as 'altruistic'. It may require considerable skill to calculate in just what circumstances particular acts will evolve. This has led some people to commit what may conveniently be called 'Sahlins' fallacy', and to suppose that the operation of kin selection requires that animals, or people, are able to perform the necessary calculations.

If this were so, kin selection could operate only in species of high intelligence. But it is not so. One of the clearest examples of kin selection occurs in a bacterial plasmid; despite misunderstandings of the phrase 'selfish gene', no one supposes that plasmids think. If, for example, members of a particular species have neighbours related to them by $r = 1/13$, say, then an act X which reduces an individual's fitness by 1 unit, and increases that of a neighbour by 14 units, will be favoured by selection. There is no need for the animal to calculate r.

Of course, if an animal *could* distinguish close relatives from distant ones, then selection would favour a gene causing altruistic acts to be directed preferentially towards the former. Can animals do this? Parents certainly care for their own offspring: is there evidence of preferential care for other relatives? There are several possible mechanisms. First, an animal might direct altruistic acts towards individuals with which it had been raised; in most cases this would have the effect of directing the acts towards relatives. Second, an animal might direct altruistic acts towards genetic relatives of those with which it was raised. This is known to happen in bees and isopods; it requires surprising powers of discrimination, and some memory. A third possibility is that an animal might be able to recognise its own genetic relatives. There is evidence that Rhesus monkeys in captivity can do this.

Animals do behave differently towards different conspecifics, both in cooperative interactions and in mate selection, and the criteria used in discrimination are correlated with actual genetic relationship. There is, however, no reason to suppose that animals have a concept of genetic relationship. In contrast, we do have such a concept. Some anthropologists have interpreted human societies on the assumption that people act so as to maximise their inclusive fitness; i.e. that they behave as predicted by kin selection theory, with the added assumption that individuals know, at least approximately, their degree of relatedness to other members of their society. I know of no very explicit discussion of how this could come about, in evolution or in individual development. The hypothesis appears to be that we have inherited from our animal ancestors the habit of discrimination, but have added an additional criterion, namely the conscious calculation of relatedness, to the criteria of propinquity, and perhaps physical and biochemical similarity, used by animals.

I turn now to evolutionary game theory. I first describe the basic ideas, to bring out the conceptual differences between classical and evolutionary game theory. I then discuss where the boundary may lie between games animals play and those that only people can play.

Imagine two animals fighting over some resource. Two 'pure' strategies are available to them – an aggressive strategy, Hawk (H), and a less risky one, Dove (D). The 'payoff matrix' might then be

$$
\begin{array}{ccc}
 & H & D \\
H & -2,-2 & 2,0 \\
D & 0,2 & 1,1.
\end{array}
$$

In evolutionary game theory, we imagine a population of animals pairing off at random and playing this game. Each animal has a strategy – pure H or pure D, or a 'mixed' strategy, 'play H with probability p and D with probability $1 - p$'. After playing, each animal reproduces its kind, and dies; the *number* of offspring produced is equal to some initial constant, say + 10, modified by the payoff received. Thus a Hawk which met a Hawk would produce 8 offspring, and one which met a Dove 12 offspring.

The population will thus evolve, and the relative frequencies of different strategies will change. The payoffs are interpreted as changes in fitness arising from the contest. Evolutionary game theory is concerned with the trajectories of evolutionary change, and in particular in finding an 'Evolutionarily Stable Strategy', or ESS. An ESS is a strategy such that, if all members of a population adopt it, no alternative, 'mutant', strategy can invade the population.

Applying this idea to the Hawk–Dove game, it is clear that H is not an ESS, because a population of Hawks would average −2 per contest, whereas a Dove mutant would average 0. Similarly, Dove is not an ESS. It turns out that the mixed strategy, 'play H with probability 1/3; play D with probability 2/3' is the only ESS of the matrix shown. A population of individuals adopting this strategy could not be invaded by any mutant. If only the pure strategists, H and D, were present, the population would evolve to a 'genetically polymorphic' equilibrium consisting of 1/3 H and 2/3 D.

Evolutionary game theory is a way of thinking about the evolution of phenotypes when fitnesses are frequency-dependent; i.e. when the best thing to do depends on what others are doing. It does not require that contests be pairwise, and is not confined to fighting behaviour; it has been applied to the evolution of the sex ratio, of dispersal, of growth strategies in plants, and so on.

As I see it, the differences between classical and evolutionary game theory concern two main points: the meaning of a 'payoff', and the contrast between dynamics and rationality, and, arising from this, the meaning of a 'solution'. Both forms of game theory require that the possible outcomes for a given player be ranked on a linear scale. In the evolutionary version, the payoffs are changes in fitness; hence, although they may be difficult to measure, they do fall naturally on a linear scale. In the classical version some difficulty arises in arranging a set of qualitatively different outcomes – e.g. loss of money, reputation, or life – on a single scale. The justification for a scale of 'utility' is that any two outcomes can always be ranked, because an individual must always have a preference between them. I shall leave to others how far this solves the difficulty; my point is that no comparable difficulty arises in evolutionary game theory.

More fundamental is the fact that evolutionary game theory is based on a well-defined dynamics – the evolution of the population – and the 'solutions' of the game are the stable stationary points of the dynamics. In contrast, classical game theory supposes rational players, and seeks a solution in terms of how such players would behave. Despite this conceptual difference, however, there is a close similarity between an ESS and a 'Nash equilibrium', which is the central equilibrium concept in classical game theory. In a two-person game, if player 1 adopts strategy A and player

2 adopts B this constitutes a Nash equilibrium if neither player would gain by changing his strategy, so long as his opponent sticks to his. An ESS differs in two respects. First, an ESS of a 'symmetric' game such as the Hawk–Dove game (i.e. a game in which there is no external asymmetry conferring different roles on the two players) requires that both players adopt the same strategy. Thus 'player 1 plays H, player 2 plays D' is a Nash equilibrium of the Hawk–Dove game, but it is not an ESS because, in the symmetric case, there is no way of distinguishing the players. Second, the definition of an ESS contains a criterion for the stability of the equilibrium which is missing from the definition of a Nash equilibrium.

Since ESS's arise from a dynamics, there is no assumption of rationality any more than there is in the case of kin selection. However, some games (or, more precisely, some strategies) do require intelligence to play. I now describe three games – the repeated prisoner's dilemma, the queuing game, and the social contract game – which are played both by men and animals, but in which the strategies available to men are more extensive than those available to animals.

An example of the prisoner's dilemma game is as follows:

		Player 2	
		Cooperate (C)	Defect (D)
Player 1	Cooperate (C)	4,4	0,5
	Defect (D)	5,0	2,2

The game is paradoxical for the following reason. No matter what player 2 does, it pays player 1 to defect. It also pays player 2 to defect. So, rationally, both should defect, yet both would be better off if they cooperated.

Not surprisingly, the only ESS of the game is Defect. Suppose, however, that the game was played between the same two opponents ten times. Consider two strategies, Defect and Tit-for-Tat (TFT) – i.e. cooperate in the first game, and subsequently do as your opponent did last time. The payoff matrix then becomes

	D	TFT
D	20	23
TFT	18	40

It is clear that TFT is an ESS; TFT strategists, when playing each other, get the benefits of cooperation. However, Defect is also an ESS. Hence there is a problem of how cooperation could evolve in the first place, although it would be stable once it had evolved. In practice, the early stages probably require the operation of kin selection. Trivers used essentially this argument to account for the evolution of 'reciprocal altruism', in which animals cooperate only with those that cooperate with them. More recently, Axelrod and Hamilton have shown (for a slightly altered model) that Tit-for-Tat is stable against any alternative strategy, and not just against Defect.

Trivers imagined that reciprocal altruism would evolve in species capable of recognising individuals and remembering how they behave, and of behaving differently towards different partners. He did not have to suppose that an animal could foresee the consequences of its behaviour; still less did he have to suppose that an animal could imagine what it would do in its partner's place. Packer has shown that baboons are capable of reciprocal altruism. Baboons have a gesture for soliciting help from others: Packer showed that those individuals which most frequently responded to solicitations from others are also those most likely to receive help when they solicit. Axelrod and Hamilton point out that reciprocal altruism could evolve without the need for individual recognition in a sessile organism; in principle, it could evolve in a plant. Thus, imagine a sessile species in which each individual has only one neighbour. Then the strategy 'cooperate with your neighbour if he cooperates; otherwise defect' would be an ESS, although I have some difficulty in seeing how it would evolve in the first place.

To summarise on the prisoner's dilemma, we must distinguish between sessile and mobile animals. For the former, since they play against only one or a few opponents, there is no need for learning; the genetically-determined strategy 'cooperate with your neighbour if he cooperates;

otherwise defect' can be stable. For reciprocal altruism in mobile animals, as demonstrated by Packer in baboons, more is needed. Since each baboon interacts with many others, and since there may be a long delay between action and reciprocation, stability requires that a baboon should recognise individuals, and remember how each has behaved, or, at the very least, associate with each individual a positive or negative sign, depending on how it has behaved. We are not, however, forced to suppose that a baboon reasons, as a man would, that it will pay to be nice to X, because X is likely to reciprocate.

A queue is a sequence of individuals, arranged according to time of arrival, and *not* according to size or strength, such that the first in the queue has prior access to some resource. Wiley reports that striped wrens form 'age queues', with the oldest male, at the head of the queue, taking over a breeding territory when the incumbent dies; such queues may be commoner than we have thought. The stability problem is clear. If a larger bird is low in the queue, why does it not displace the bird at the head? We do not know the answer, although there are several possibilities.

My reason for mentioning queues, however, is that there are ways in which a human queue might be stabilised which are unlikely to operate among animals. Some possibilities are as follows:

  (i) A man stays in line because he has been taught that it is wrong not to.
 (ii) A man who jumps the queue will acquire a reputation which will damage him in later social contacts.
(iii) A man who attempts to jump the queue will be restrained by the police.
 (iv) Any attempt to jump the queue will be resisted by all other members.

The first three of these possibilities explain stability by events external to the queue itself. If we take reciprocal altruism seriously, method (ii) might conceivably operate in animals, although not in the case of age queues; methods (i) and (iii) could not. Method (iv) is the most interesting.

Clearly, collective resistance could stabilise the queue; the problem is why individuals should join in collective resistance. Let us consider animals first. It is conceivable that collective resistance would be individually advantageous. For example, suppose that, in the queue $\alpha - \beta - \gamma$, α is challenged by γ. It might pay β to help α to beat off the challenge, because if he does not the sequence might become $\gamma - \alpha - \beta$, and β has dropped from second to third. (It is harder to see why γ should help α to resist a challenge by β.) If the alternative strategies are 'resist only challenges to oneself', and 'resist any challenge', the latter would be favoured by selection provided that it was usually advantageous. It would not have to be advantageous in every possible case. Thus I can imagine that animal queues are stabilised by collective resistance, without having to suppose that individual animals perform complex calculations. However, there is no evidence that this is what happens.

In human beings, some individuals might perceive that it would be in the general interest if queue-jumping were prevented, particularly if a complete breakdown of the queue made the resource unavailable (if the driver sees a fight at a bus stop, he doesn't stop). If so, they might persuade the queue members to bind themselves to wait in line, and to punish transgressors. Note that this is possible even if there are a minority of individuals (perhaps the strongest person in the queue) who do not benefit. This brings me to my third game, the 'social contract' game.

Suppose that the payoff to all the members of a small group is greater if all cooperate than if all defect. Then the members might agree to cooperate, and to join in punishing any member who defects. Even if the act of punishing was cheap, and of being punished expensive, this would still not be sufficient to guarantee stability, because of the problem of the 'free rider'. Thus a member who cooperated, but did not join in punishing, would be better off than someone who cooperated and did join in punishing. Hence a social contract can ensure stable cooperation only if it reads 'I will cooperate; I will join in punishing any defection; I will treat any member who does not join in punishing as a defector'.

Could an analogous behaviour occur in animals? If, for

some specific action X (e.g. jumping an age queue), animals
(i) did not do X, (ii) drove out of the group any animal that
did X, and (iii) drove out any animal that did not join in
driving out an X-doer, then the behaviour would be
evolutionarily stable. The difficulty, of course, lies in
imagining how such a complex behavioural syndrome, which
is stable only when complete, could arise in the first place.
This illustrates Elster's point that natural selection is a hill-
climbing process which can only reach local optima, whereas
rational behaviour can reach a global optimum. (Elster is,
however, wrong in thinking that natural selection cannot
reach the mixed ESS of the Hawk–Dove game.)

Despite the difficulty of imagining how a behaviour
involving the three components outlined in the last paragraph
could arise in the first place, I think it is quite possible that
the explanation of stable age queues in animals may be of
this kind. A similar mechanism may perhaps account for the
fact that some group-living animals drive sick or injured
individuals out of the group. To drive out a sick individual
is sometimes advantageous, because the sickness may be
infectious. If, in the above specification, we replace 'doing
X' by 'being different from typical members of the group',
we have a mechanism that will explain this behaviour.

The crucial difference between men and animals, then, lies
in the nature of the action X which is proscribed by the
contract. In animals, X would have to be genetically specified,
although it might be specified merely as being different, in
any way, from other members of the species. In man, X
could be a newly acquired possibility (e.g. human cloning
or hang-gliding), perceived by some individuals as being
socially undesirable, the perception being communicated to
others linguistically. Even language may not be enough to
account for the agreement in a social contract not to do X,
when X is not genetically specified. Thus, suppose that a
member of the group recognises that he or she would be
better off if no one did X. Before that member would
embark on an attempt to persuade others, he would have to
recognise that others might feel about X as he did.

To play the social contract game successfully, therefore,
when the prescriptions of the contract are culturally rather

than genetically specified, an animal would have to think of others as having motivations similar to its own, so that it could foresee their future behaviour, and it would have to communicate symbolically. The game is therefore useful in illustrating the kinds of strategies animals cannot adopt. However, I do not think it is a particularly appropriate model of human social interactions, for two reasons. First, it treats all members of a social group as having the same set of possible actions. In fact, owners of land or factories can do things non-owners cannot, as can men with weapons compared to those without, or even men as compared to women. Hence social contracts may bind, not all members of a society, but members of some group within society. The problem is to explain how individuals come to identify their interest with that of a specific group, and why different societies tend to divide into groups along different lines, according to economic class, religion, race, etc.

The other weakness of the social contract model lies in its excessively rational and legalistic nature. In practice, I suspect that 'contracts' are arrived at as much by religious and ideological persuasion as by rational discourse, and maintained more by the threat of social ostracism than by legal restraint. The ideological nature of social contracts means that they need not always correspond to individual self-interest. However, they cannot depart too far from it; men can be swayed by beliefs, but not too far. Of course, one group in society may be able to impose its will on another.

Returning to my brief of animal intelligence, animals can and do act as members of a group against other conspecific groups. The groups are usually composed of genetic kin, but not always. The cohesion of the group may be cemented by joint display activities, as in the cacophony of a troop of howler monkeys. However, I would not claim that these displays have any culturally mediated symbolic meaning, as do the myths and rituals which bind human groups. Perhaps the main consequences of the lack of high intelligence in animals is that they are not as good as we are at fooling themselves.

22

Evolution and the Theory
of Games

I want in this article to trace the history of an idea. It is beginning to become clear that a range of problems in evolution theory can most appropriately be attacked by a modification of the theory of games, a branch of mathematics first formulated by Von Neumann and Morgenstern in 1944 for the analysis of human conflicts. The problems are diverse and include not only the behaviour of animals in contest situations but also some problems in the evolution of genetic mechanisms and in the evolution of ecosystems. It is not, however, sufficient to take over the theory as it has been developed in sociology and apply it to evolution. In sociology, and in economics, it is supposed that each contestant works out by reasoning the best strategy to adopt, assuming that his opponents are equally guided by reason. This leads to the concept of a 'minimax' strategy, in which a contestant behaves in such a way as to minimise his losses on the assumption that his opponent behaves so as to maximise them. Clearly, this would not be a valid approach to animal conflicts. A new concept has to be introduced, the concept of an 'evolutionarily stable strategy'. It is the history of this concept I want to discuss.

Evolution of the sex ratio

Consider first the evolution of the sex ratio. In most animals and plants with separate sexes, approximately equal numbers of males and females are produced. Why should this be so? Two main kinds of answer have been offered. One is couched in terms of advantage to the population. It is argued that the sex ratio will evolve so as to maximise the number of meetings between individuals of opposite sex. This is essentially a 'group selection' argument. The other, and in my view certainly correct, type of answer was first put foward by Fisher. It starts from the assumption that genes can influence the relative numbers of male and female offspring produced by an individual carrying the genes. That sex ratio will be favoured which maximises the number of descendants the individual will have and hence the number of gene copies transmitted. Suppose that the population consisted mostly of females: then an individual which produced only sons would have more grandchildren. In contrast, if the population consisted mostly of males, it would pay to have daughters. If, however, the population consisted of equal numbers of males and females, sons and daughters would be equally valuable. Thus a 1:1 sex ratio is the only stable ratio; it is an 'evolutionarily stable strategy'.

Fisher allowed for the fact that the cost of sons and daughters may be different, so that a parent might have a choice, say, between having one daughter or two sons. He concluded that a parent should allocate equal resources to sons and daughters. Although Fisher wrote before the theory of games had been developed, his theory does incorporate the essential feature of a game – that the best strategy to adopt depends on what others are doing. Since that time, it has been realised that genes can sometimes influence the chromosome or gamete in which they find themselves, so as to make that gamete more likely to participate in fertilisation. If such a gene occurs on a sex-determining (X or Y) chromosome, then highly aberrant sex ratios can evolve.

More immediately relevant are the strange sex ratios in

certain parasitic hymenoptera (wasps and ichneumonids). In this group of insects, fertilised eggs develop into females and unfertilised eggs into males. A female stores sperm and can determine the sex of each egg she lays by fertilising it or leaving it unfertilised. By Fisher's argument, it should still pay a female to produce equal numbers of sons and daughters. More precisely, it can be shown that if genes affect the strategy adopted by the female, then a 1:1 sex ratio will evolve. Some parasitic wasps lay their eggs in the larvae of other insects, and the eggs develop within their host. When adult wasps emerge, they mate immediately before dispersal. Such species often have a big excess of females. This situation was analysed by Hamilton. Clearly, if only one female lays eggs in any given larva, it would pay her to produce one male only, since this one male could fertilise all his sisters on emergence. Things get more complicated if a single host larva is found by two parasitic females, but the details of the analysis do not concern us. The important point is that Hamilton looked for an 'unbeatable strategy' – that is, a sex ratio would be evolutionarily stable. In effect, he used Fisher's approach but went a step farther in recognising that he was looking for a 'strategy' in the sense in which that word is used by game theorists.

Animal contests and game theory

A very similar idea was used by the late G. R. Price in an analysis of animal behaviour. Price was puzzled by the evolution of ritualised behaviour in animal contests – that is, by the fact that an animal engaged in a contest for some valuable resource does not always use its weapons in the most effective way. Examples of such behaviour have been discussed by Lorenz, Huxley, and others. It seems likely that ethologists have underestimated the frequency and importance of escalated, all-out contests between animals. Yet display and convention are a common enough feature of animal contests to call for some explanation. Both Lorenz and Huxley accepted group selectionist explanations: Huxley, for example, argued that escalated contests would result in many animals being seriously injured, and 'this would militate against the survival of the species'. Similar assumptions are widespread in ethology, although not often so clearly expressed.

Price was reluctant to accept a group selection explanation. It occurred to him that if animals adopted a strategy of 'retaliation', in which an animal normally adopts conventional tactics but responds to an escalated attack by escalating in return, this might be favoured by selection at the individual level. He submitted a paper to *Nature* arguing this point, which was sent to me to referee. Unfortunately the paper was some fifty pages in length and hence quite unsuitable for *Nature*. I wrote a report saying that the paper contained an interesting idea, and that the author should be urged to submit a short account of it to *Nature* and/or to submit the existing manuscript to a more suitable journal. I then thought no more of the matter until, about a year later, I spent three months visiting the department of theoretical biology at Chicago. I decided to spend the visit learning something about the theory of games, with a view to developing Price's idea in a more general form and applying it to certain other problems. I was at that time familiar with Hamilton's work on the sex ratio (indeed, the work formed part of his PhD thesis, of which I was the external examiner), but I had not seen its relevance to Price's problem.

While at Chicago, I developed the formal definition of an evolutionarily stable strategy which I will give in a moment and applied it to the 'Dove–Hawk–Retaliator' and 'War of Attrition' games. I also realised the similarity between these ideas and the work of Hamilton (and also MacArthur) on the sex ratio. When I came to write up this work, it was clearly necessary to quote Price. I was somewhat taken aback to discover that he had never published his idea and was now working on something else. When I returned to London I contacted him, and ultimately we published a joint paper in which the concept of an evolutionarily stable strategy was applied to animal contests.

At this point it will be convenient to describe some ideas from the theory of games. By 'game' or 'contest' is meant an encounter between two individuals (I am not concerned with n-person games) in which the various possible outcomes would not be placed in the same order of preference by the two participants: there is a conflict of interest. By 'strategy' is meant a complete specification of what a contestant will do in every situation in which he might find himself. A strategy may be 'pure' or 'mixed'; a pure strategy states 'in situation A, always do X': a mixed strategy states 'in situation A, do X with probability P and Y with probability Q'. Suppose that there are three possible strategies, A, B and C. A 'payoff matrix' is then a 3×3 matrix listing the expected gains to a contestant adopting these three strategies, given that his opponent adopts one of the other strategies.

These ideas will be made clearer by an example. Consider the children's game 'Rock–Scissors–Paper'. In each contest, a player must adopt one of these 'strategies' in advance: then Rock blunts Scissors, Scissors cuts Paper, and Paper wraps Rock. Suppose that the winner of each contest receives one dollar from the loser; if both adopt the same strategy, no money changes hands. The payoff matrix is:

	R	S	P
R	0	+1	−1
S	−1	0	+1
P	+1	−1	0

The payoffs are to the player on the left. This particular game is a 'zero-sum' game, in the sense that what one player wins the other loses; in general the games considered below are not of this kind. Clearly, a player adopting the pure strategy 'Rock' will lose in the long run, because his opponent will catch on and play 'Paper'. A player adopting the mixed strategy '1/3 Rock, 1/3 Scissors, 1/3 Paper' will break even.

How are these ideas to be applied to animal contests? A genotype determines the strategy, pure or mixed, that an animal will adopt. Suppose an animal adopts strategy I and his opponent strategy J; then the payoff to I will be written $E(I,J)$ where E stands for 'expected gain'. This payoff is the change in I's fitness as a result of the contest, fitness being the contribution to future generations.

Evolutionarily stable strategy

We are now in a position to define an evolutionarily stable strategy, or ESS for short. Suppose that a population consists of individuals adopting strategies I or J with frequencies p and q, where $p + q = 1$. What is the fitness of an individual adopting strategy I?

$$\text{Fitness of } I = p.\,E(I,I) + q.E(I,J)$$
$$\text{Fitness of } J = p.E(J,I) + q.E(J,J)$$

If a particular strategy, say I, is to be an ESS, it must have the following property. A population of individuals playing I must be 'protected' against invasion by any mutant strategy, say J. That is, when I is common, it must be fitter than any mutant. That is, I is an ESS if, for all $J \neq I$,

$$either\ E(I,I) > E(J,I)$$
$$or\ E(I,I) = E(J,I)$$
$$and\ E(I,J) > E(J,J). \tag{1}$$

If these conditions are satisfied, then a population of individuals playing I is stable; no mutant can establish itself in such a population. This follows from the fact that when q is small, the fitness of I is greater than the fitness of J.

It is important to emphasise at this point that the ESS is not necessarily the same as the strategy prescribed by game theorists for human players. There the assumption is that a player will adopt that strategy which minimises his losses, given that his opponent plays so as to maximise them. Lewontin applied such 'minimax' strategies to evolution. He was concerned with a contest not between individuals but between a species and 'nature'. The objective of a species is to survive as a species – to avoid extinction. It should therefore adopt that strategy which minimises its chances of extinction, even if nature does its worst. That is, the species must adopt the minimax strategy. For example, a species should retain sexual reproduction rather than parthenogenesis, because this will enable it to evolve to meet environmental change. This is clearly a group selectionist approach; the advantage is to the species and not to the individual female. In contrast, the concept of an evolutionarily stable strategy is relevant to contests between individuals, not between a species and nature, and is concerned solely with individual advantage.

Let us now apply these ideas to a particular problem. Suppose that two animals are engaged in a contest for some indivisible resource which is worth $+V$ to the victor. An animal can 'display', or it can 'escalate' – in which case it may seriously injure its opponent – or it can retreat, leaving its opponent the victor. Serious injury reduces fitness by $-W$ (a 'wound') and forces an animal to retreat. Finally, a long contest costs both animals $-T$. The two simplest strategies are

Hawk. Escalate, and continue to do so until injured or until opponent retreats.
Dove. Display. Retreat if opponent escalates, before getting injured.

We suppose that two Hawks are equally likely to be injured or to win. We also suppose that two Doves are equally likely to win, but only after a long contest costing both of them $-T$. The payoff matrix is then

	H	D
H	$\frac{1}{2}(V-W)$	V
D	0	$\frac{1}{2}V-T$

Since arithmetic is easier than algebra, let us take $W = 6$, $V = 4$, and $T = 1$; these values are arbitrary, but will illustrate the logic of the game. The matrix becomes:

	H	D
H	-1	4
D	0	1

Clearly, there is no pure ESS. Thus H is not an ESS,

because in a population of Hawks, a typical H individual has a payoff of -1, and a D mutant of 0: therefore D can invade a population of Hawks. A similar argument shows that H can invade a population of Doves. For these particular values, the only ESS is 'Play H with probability 3/4; play D with probability $\frac{1}{4}$'. The probabilities will change with the values of W, V and T, but, provided $W > V$, the ESS will be a mixed one. At an evolutionary equilibrium, the population will consist of individuals that play sometimes Hawk, and sometimes Dove: alternatively, it may consist of some individuals that always play Hawk, and some that always play Dove.

Price's suggestion was that a third strategy, Retaliator, R might be an ESS. R plays D against D, and H against H. The payoff matrix is

	H	D	R
H	-1	4	-1
D	0	1	1
R	-1	1	1

It turns out that the situation is a bit more complicated than Price imagined. A population of Retaliators is stable against invasion by H, but Dove mutants would have the same fitness as Retaliators, and so might drift up in frequency. However, if the game is modified in plausible ways, Retaliator is indeed an ESS.

This analysis suggests that we would expect to find retaliation a feature of actual behaviour. One example must suffice: a rhesus monkey which loses a fight will passively accept incisor bites but will retaliate viciously if the winner uses its canines.

One assumption made above – that two Doves can settle a contest – needs some justification. Why don't they go on forever? Consider the following game. Two players, A and B, can only display. The winner is the one who goes on for longer; the only choice of strategy is how long to go on for. A selects time T_A and B selects T_B. The longer the contest actually continues, the more it costs the players; the

costs associated with these times are m_A and m_B. If $T_A >$
T_B, then we have

$$\text{payoff to } A = V - m_B$$
$$\text{payoff to } B = -m_B$$

The cost of m_A which A was prepared to pay, is irrelevant,
provided that it is greater than m_B. Our problem then is:
How should a player choose a time, and a corresponding
value of m? More precisely, what choice of m is an ESS?
For obvious reasons, I have called this the 'War of Attrition'.
Clearly, no pure strategy can be an ESS. Any population
playing m, say, could be invaded by a mutant playing M,
where $M > m$; if $m > V/2$, it could also be invaded by a
mutant playing 0. It can be shown that there is a mixed ESS
given by

$$p(x) = \frac{1}{V} e^{-x/V} \qquad (2)$$

where $p(x)\delta x$ is the probability of playing m between x and
$x + \delta x$.

What does this mean? There are two possible ways in
which an ESS of this kind could be realised. First, all
members of the population might be genetically identical
and have a behaviour pattern which varied from contest to
contest according to Eq. 2. Second, the population might
be genetically variable, with each individual having a fixed
behaviour, the frequencies of different kinds of individuals
being given by Eq. 2. In either case, the population would
be at an ESS.

G. A. Parker has described a situation which agrees rather
well with Eq. 2. Female dung flies of the genus *Scatophaga*
lay their eggs in cow pats. The males stay close to the cow
pats, mating with females as they arrive to lay their eggs.
What strategy should a male adopt? Should he stay with a
pat once he has found one, or should he move on in search
of a fresh pat as soon as the first one begins to grow stale?
This is comparable to the choice of a value of m in the War

of Attrition. His choice will be influenced by the fact that females arrive less frequently as a pat becomes staler. His best strategy will depend on what other males are doing. Thus if other males leave a pat quickly, it would pay him to stay on, because he would be certain of mating any females which do come. If other males stay on, it would pay him to move.

Parker found that the actual length of time males stayed was given by a distribution resembling Eq. 2. By itself this means little, because it is the distribution one would expect if every male had the same constant probability of leaving per unit time. It is the typical negative exponential distribution expected for the 'survivors' of a population suffering a constant 'force of mortality'. What is significant is that Parker was able to show that the expected number of matings was the same for males which left early as for those which stayed on. This means that the males are adopting an ESS; natural selection has adjusted the probability of leaving per unit time to bring this about.

It is not known whether contests between pairs of animals, in which only display is employed, show the appropriate variation in length. It will be interesting to find out.

A major complication in applying these ideas in practice arises because most contests are asymmetrical either in the fighting ability of the contestants (i.e. in what Parker has called 'resource-holding potential' or RHP), or in the value of the resource to the contestants (i.e. in payoff). Clearly, these asymmetries can only affect the strategies adopted if they are known to the contestants. Thus suppose two animals differ in size, and hence in RHP, but have no way short of escalation of detecting the difference. Then the difference cannot alter their willingness to escalate (i.e. their strategy), although it would affect the outcome of an escalated contest.

In some cases an asymmetry may be clearly perceived by both contestants but have relatively little effect on RHP or on payoff. The obvious example is the asymmetry between the 'owner' of a resource (e.g. a territory, a female, an item of food) and a 'latecomer'. There is no general reason why an owner should have a higher RHP than a latecomer. The value of a resource will often be greater to the owner, but, as I shall show in a moment, no such difference is necessary before an asymmetry can be used to settle a contest conventionally.

To fix ideas, consider the numerical example of the game of Hawks and Doves discussed above. Suppose that an animal may be either the owner of a resource, or an interloper, and that a particular animal is equally likely to find itself playing either role. Consider the strategy: 'Play H if you are owner: play D if you are interloper'. I have called such a strategy 'Bourgeois', or B for short. The payoff matrix then becomes

	H	D	B
H	-1	4	1.5
D	0	1	0.5
B	-0.5	2.5	2

Then B satisfies the ESS conditions against either H or D. Thus the Bourgeois strategy, which amounts to the conventional acceptance of ownership, is an ESS against any strategy which ignores ownership.

It follows that conventional acceptance of ownership can be used to settle contests even when there is no asymmetry in payoff or RHP, provided that ownership is unambiguous. Some actual examples will help to illustrate this pont.

The ESS in practice

The hamadryas baboon, *Papio hamadryas*, lives in troops composed of a number of 'one-male groups', each consisting of an adult male, one or more females, and their babies. The male, who is substantially larger than the females, prevents 'his' females from wandering away from his immediate vicinity; a female rapidly comes to recognise this 'ownership'. It is rare for an owning male to be challenged by another. How is this state of affairs maintained?

Kummer describes the following experiment. Two males, previously unknown to each other, were placed in an enclosure; male A was free to move about the enclosure whereas male B was shut in a cage from which he could see what was happening but not interfere. A female strange to both males was then loosed into the enclosure. Within 20 minutes male A had convinced the female of his ownership, so that she followed him about. Male B was then released into the enclosure. He did not challenge male A, but kept well away from him, accepting A's ownership.

These observations can be explained in two ways. First, male B may have been able to detect that male A would win an escalated contest if challenged; second, there may be a conventional acceptance of ownership, for the reasons outlined above. Kummer was able to show that the second explanation is correct. Two weeks later, he repeated the experiment with the same two males but with a different female, but on this occasion male B was loose in the enclosure and male A confined. Male B established ownership of the female and was not challenged by A.

One last observation is relevant. If a male is removed from a troop, his females will be taken over by other males. If after some weeks the original male is reintroduced, an escalated fight occurs; both males now behave as 'owners'.

It could be argued that in the hamadryas baboon there is a difference in payoff, because when a male first takes over a new female he has to invest time and energy in persuading her to accept his ownership. This is probably correct, although the theoretical analysis shows that no such difference is required for the establishment of an ESS based on conventional acceptance of ownership. An asymmetry in payoff is less likely in the anubis baboon studied by Packer. In this species, there is a fairly stable male dominance hierarchy for food but not for females. Females are not the permanent property of particular males; instead, a male 'owns' a female only for a single day – or for several days if he can prevent her from moving away during the night. Once in temporary possession of a female, a male is not challenged, even by those above him in the dominance hierarchy. Why should contests about food and females be settled differently? One possible explanation is that the ownership principle could not be used to settle contests over food, because it must often be the case that two animals see a food item almost simultaneously. Ownership would be ambiguous; two animals would both regard themselves as owners of the same item, and escalated contests would ensue.

This last possibility is beautifully illustrated by the work of L. Gilbert on the swallowtail butterfly, *Papilio zelicaon*. Because this is a relatively rare butterfly, the finding of a sexual partner presents a problem. This problem is solved by 'hilltopping'. Males establish territories at or near the tops of hills, and virgin females fly uphill to mate. There are, however, more males than hilltops, so most males must accept territories lower down the slopes. They attempt to waylay females on their way up and, although they sometimes succeed, the evidence suggests that the male actually at the hilltop mates most often. Gilbert marked individual males and observed that a strange male did occasionally arrive at a hilltop and challenge the owner, but the stranger invariably retreated after a brief 'contest'.

As in Kummer's experiments with baboons, we have to choose between two explanations. Either the owner of a hilltop is a particularly strong butterfly, and this fact is perceived by the challenger during a brief contest, or there is again an ESS based on conventional acceptance of ownership. Gilbert showed that the latter explanation is correct by an experiment analogous to that of removing a male hamadryas baboon from a troop and then restoring it. He allowed two male butterflies to occupy a hilltop on alternate days, keeping each in the dark on their off days. After two weeks, when both males had come to regard themselves as owners of the same hilltop, he released them on the same day. A contest lasting many minutes and causing damage to the contestants ensued.

Much has been left out of these simple models. Contests in which only partial information about asymmetries is available to the contestants, or in which information is acquired in the course of a contest, are discussed by Maynard Smith and Parker. The same paper discusses the possibility of 'bluff' – that is, the possession of structures such as manes, ruffs or crests, which increase apparent RHP without an equivalent increase in actual fighting ability.

I suggested at the beginning of this article that the concept of an ESS is also relevant to the evolution of ecosystems; this idea is developed by Maynard Smith and Lawlor (in press). It is impossible to do more here than indicate the nature of the problem. In nature, animals and plants compete for resources – food, space, light, etc. Genetic changes in an individual can alter its 'choice' of resources: for example, the food items taken by an animal, or the time of year a plant puts out its leaves. Individuals will choose their resources so as to maximise their fitness. The best choice will depend on what other individuals, of the same and other species, are doing. If everyone else is eating spinach it will pay to concentrate on cabbage; since most forest trees put out their leaves late in spring, it pays forest herbs to put out leaves early.

Since the appropriate strategy for an individual depends on what others are doing, we are again concerned with the search for an ESS. Lawlor and I conclude that two competing species will tend to become specialists on different resources, even though in isolation each species would be a generalist. This conclusion is not a new one: it accords with a good deal of observational data and has received several previous theoretical treatments. We would claim, however, that we have clarified a familiar idea and set it in a wider context. That wider context is simply this: whenever the best strategy for an individual depends on what others are doing, the strategy actually adopted will be an ESS.

PART 5

THE LAWS OF THE GAME

The essays in this section deal with topics as various as the origin of life, the development of segmentation in animals, and the perception of distance. What they have in common is that, although they contain no mathematics, they look at the world from the perspective of a mathematician.

One mathematical perspective has influenced me more than any other. It is the belief that the behaviour of things depends on their structure, and on the relations between their parts. Systems with similar structures will behave similarly, whether their parts are molecules, cells, transistors or animals. The language we use to describe these structures and relations is mathematics. In different ways, *The Counting Problem, Understanding Science*, and *Matchsticks, Brains and Curtain Rings*, illustrate this approach.

Hypercycles and the Origin of Life is more technical than the other essays in this book. It was originally written as an attempt to introduce the highly mathematical ideas of Manfred Eigen and Peter Schuster to molecular biologists, who like to pretend that they know no mathematics. I therefore got rid of the formal mathematics, but assumed a knowledge of molecular biology. The essay will therefore be difficult for readers ignorant of molecular biology, but I have left it in, because the ideas are interesting enough to be worth struggling with.

Popper's World, and *Rottenness is all*, treat the old problem of the relation between determinacy and free will. I do not expect anyone to agree with me about this.

Finally, I have a query. In Matchsticks, Brains and Curtain Rings, I describe a solution to one version of the classic Travelling Salesman problem. I did not think of it myself, but I cannot remember who told it me, or where I read it. Is there anyone out there who knows?

23

The Counting Problem

Development seems to occur in a stepwise fashion, the completion of each step setting the stage for the next. Each step may be comparatively simple, even if the end result of a long sequence of such steps is exceedingly complicated. One way in which a theoretical approach may be helpful is in identifying characteristic steps (of which there may be comparatively few kinds), and in suggesting possible mechanisms for them. This is the approach adopted in this paper; it was also adopted by Wolpert in his contribution to the conference.

The particular developmental step considered is the generation of a constant number of similar parts. For example, how is it that most men develop five fingers on each hand and 29 precaudal vertebrae? This problem, which I shall call the counting problem, resembles Wolpert's problem of the French Flag in that it illustrates in a simple form many of the characteristics of morphogenesis. In discussing it I shall first discuss what kinds of 'counting machine' could exist, and then ask to which kind embryos belong.

By a 'counting machine' I mean a machine which can do one of two things. It can either do something a fixed number

N times when it is stimulated, or do something when it has been stimulated N times. A reversible machine which could do one of these things could do both, and even for irreversible machines there is likely to be a structural similarity between the two types. Embryos are of the former type; strictly, a single egg only develops five fingers on the right hand once, but a set of similar eggs each develops five fingers, which is equivalent to a single egg developing five fingers every time it is fertilised. Animals with nervous systems may be of either type, i.e. they may perform an action when they have been stimulated a fixed number of times, or they may respond a fixed number of times to a single stimulus.

Strictly, such machines ought to be called 'number generating machines', and the phrase 'counting machines' should be reserved for those which can do something a variable number of times (the number varying predictably according to the nature of the stimulus received) or which respond in qualitatively different ways according to the number of times they have been stimulated.

There would seem to be two main kinds of counting machine, which I will call ratio counters and digital counters. A simple ratio counter is shown in Figure 23.1.

Suppose that the volume of water required to fill the upper vessel is n times that required to fill the lower one from A to B, and that when the lower vessel is almost full it tips the seesaw. Then if n lies between 5 and 6, the seesaw will rock five times if the upper vessel is filled. There are two points to note about this machine, which appear to be general properties of ratio counters. First, there has to be some 'quantising mechanism' – in this case supplied by a syphon with a flow rate which is rapid compared to the rate of flow from the upper vessel – which will break up a continuous process into a number of discrete events. Second, the number n generated depends on the ratio between two continuous variables; if n is to be constant, then the range of variation of these variables must be restricted.

In digital counting, the 'machine' must again be able to generate discrete events, but the number of events is regulated by pairing them off one by one with a pre-existing set of entities contained in the machine. These entities may be all

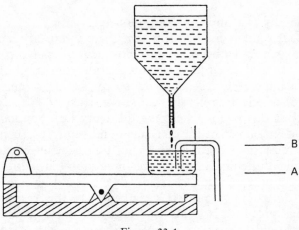

Figure 23.1

alike. For example, in cricket a bowler delivers six successive balls and is then replaced by a second bowler. The umpire often counts these deliveries by starting with six pebbles in his left-hand pocket, and after each delivery passing one pebble to his right-hand pocket. When he has moved the last pebble, he calls 'over', and the second bowler takes over.

In other cases the pre-existing entities may be unalike, and connected in sequence by rules. Thus, an umpire who wished to place six pebbles in his pocket might take pebbles one at a time from a pile and place them in his pocket, saying as he did so, 'one – two – three – four – five – six'. To do this he must have in his brain before he starts 'representations' of six qualitatively different numbers, together with a set of counting rules (e.g. after saying four, say five) and a rule saying: 'When you have said six, stop.' This is a mechanism which can easily be modified (by varying the last rule) to generate different numbers in response to different inputs, i.e. it is a counting machine proper rather than merely a number-generator.

Notice that in digital counting the counting machinery is separate from the quantising mechanism; thus in the first example the umpire with his pebbles is the counting machine, but he is not responsible for the fact that bowlers deliver balls one by one and not continuously.

At first sight it might seem that there is no difficulty in explaining how embryos count. For example, an organism which required to produce a polypeptide consisting of 163 phenylalanine residues could do so by having a length of DNA of which the relevant strand consisted of 489 adenine bases. But this is not where the difficulty lies. Suppose for example an animal had a constant number of 163 segments (to the best of my knowledge, no animal achieves such a feat), the difficulty lies in explaining how the ability to produce a polypeptide 163 residues long could be translated during development into 163 segments. In this respect the counting problem is typical of morphogenetic problems; thus molecular biology is able to explain the shape of macromolecules – or it will be able to do so when the way in which amino acid sequence determines tertiary structure is understood. But it does not seem very likely that the *shapes* of cells or of higher organisms are determined by the shapes of their constituent molecules, as the shape of jigsaw is determined by the shape of its pieces, or the shape of a virus may be determined by the shape of its constituent molecules.

Mechanisms of ratio counting in embryos are not difficult to think of. One possible mechanism was suggested by Turing. Normally, when a number of reacting chemical substances are free to diffuse throughout some region, or 'morphogenetic field', they will reach a stable equilibrium distribution such that their concentrations are uniform over the field. Turing showed that for certain values of the reaction and diffusion rates this uniform equilibrium is unstable, and any small disturbance will lead to a standing wave of concentrations, with a 'chemical wavelength' λ separating the peaks of concentration. If one of these substances induced cells exposed to it to differentiate in a particular way – e.g. to form a bristle – such a process could account for the regular spacing of structures on a surface. This mechanism can be made somewhat more plausible by allowing for recent knowledge about the control of protein synthesis.

The development of a chemical wave provides a quantising

mechanism. If a wave develops within a field of given size, a counting mechanism is also provided. Thus in the one-dimensional case, if waves of chemical wavelength λ develop in a field of length s, then, since an integral number of waves must be formed, the number actually formed is typically the nearest integer to s/λ.

If for some reason the size of the morphogenetic field is changed, then we should expect a change in the number of structures formed. This is often the case – for example, Kroeger has described how the number of wing veins in *Ephestia* can be altered by altering the size of the field. Nevertheless, difficulties do arise in trying to explain the constancy of numbers of repeated parts by ratio counting. For example, if the number 30 is to be generated, with an error rate of 5 per cent or less (and vertebrates, annelids and arthropods all do as well as this in determining segment number), then the ratio s/λ would have to have a coefficient of variation of less than 1 per cent. A review of the coefficient of variation of some macroscopic linear dimensions in animals suggests that 5 per cent is a low value. Thus the accuracy with which segment number is determined is something of a paradox. It does not really help to say that dimensions on a molecular scale can be determined very accurately, since this leaves a problem of translation.

One possible solution is a process of multiplication. Suppose that instead of generating the number 30 in one step, the body were first divided into five parts, and each of these were then divided into six parts; the total number 30 could be accurately generated without excessive accuracy in the individual steps. There is a simple way of spotting this if it happens. Variants, if they do occur, will differ from the norm not by one but by some larger number. It is interesting therefore that the segment number in centipedes varies in twos. But I know of little other evidence for processes of multiplication. Perhaps the most illuminating feature of the 'multiplication' model is that it provides what I believe is the most important reason why development is stepwise and hierarchical: there is a limit to the complexity of the patterns which can reliably be generated in one step.

Digital counting may be involved when a number shows

great constancy both in evolution and development. Some examples of digital counting are trivial. For example, the number of spinal nerves is determined by the number of somites by digital counting. But this is uninteresting, mainly because it does not explain how microscopic information is translated into macroscopic, i.e. how the specification of macromolecules specifies the large-scale structure. A more promising type of digital counting is that using qualitatively different entities. In the process of embryonic induction, a tissue of kind A causes a second tissue with which it comes into contact to differentiate into kind B. This is exactly comparable to a 'rule of counting', whereby each number determines the characteristics of the next number.

Experiments suggesting a digital counting mechanism were performed by Moment on the worm *Clymenella*, which has a constant number of 22 segments. If a few segments are removed simultaneously from the anterior and posterior end of the worm, it can regenerate back to the original number of 22. More relevant, the isolated piece retains its position in the series; thus if for example four segments are removed anteriorly and two posteriorly, exactly those numbers are regenerated.

These observations are inexplicable unless there is some difference between the segments. It is interesting though that *Clymenella* can count backwards as well as forwards. The simplest rule for this would be: 'If a C is in contact with a B, it causes the next segment to be a D; if a C is in contact with a D, it causes the next segment to be a B.' This brings us very close to Wolpert's problem of the French Flag.

24

Understanding Science

This is a translation of a book first published under the title *Das Spiel* in 1975. It is an ambitious book whose aim is to convey to the reader what it is to have a well-furnished scientific mind. Some years back, C. P. Snow persuaded us that the diagnostic characteristic of such a mind is familiarity with the second law of thermodynamics. His particular choice of a scientific law was unfortunate, because it is easier to talk nonsense about the second law than almost anything else, but in principle he was on the right track. A knowledge of theories is more relevant than a knowledge of facts. Biologists have to know a lot of facts, while physicists seem to know almost nothing. But although it is true that a well-educated scientist will be familiar with a number of theories, from Newton's laws to the central dogma of molecular biology, I do not think that this is the critical distinction between understanding science and not understanding it. I suggest, instead, that it is a familiarity with the ways in which systems with different structures and relationships are likely to behave. It is this familiarity that Eigen and Winkler try to convey.

I can best explain this by giving an example. One of the things I learnt when I was an aircraft engineer during the war was that if a control system has a time delay in the feedback loop – that is, if some time elapses between a control action and its effects on the object to be controlled – then the system is likely to oscillate. Consequently, whenever I come across a system which is oscillating, whether it be the menstrual cycle or the numbers of hares and lynxes in Canada, I look for delayed feedback. In doing so, I am assuming that structure determines behaviour. That is, if the components of a system are related to one another in particular ways, then the system will behave accordingly. The behaviour is determined by the structure, and not by whether the components are electrical circuits, hormones or animals.

What a scientist has to acquire, then, is not merely a knowledge of specific theories relevant to his particular field, but an understanding of how the behaviour of a system depends on the relationship between its parts. Typically, the behaviour is described mathematically. One learns that simple harmonic motion is described by a particular differential equation, whether the variable that is behaving harmonically is the current in a tuning circuit or the velocity of a weight suspended from a spring. The major difficulty in communicating scientific ideas to a non-professional audience, therefore, is the lack of a common mathematical language.

Eigen and Winkler attempt to bridge this language gap. The concepts they aim to explain are not the relatively simple engineering examples of delayed feedback and harmonic motion. Instead, they discuss many of the most fundamental and difficult ideas in contemporary science. To give a list, which may be meaningful only to scientists, they discuss the concepts of equilibrium, competition, natural selection, entropy and information, symmetry and group theory, pattern formation, and conservative and dissipative structures. At the same time, they apply these concepts to different levels of organisation, from chemistry through development, ecology and evolution to economics and music.

Despite this vast range of topics, the authors avoid the use of mathematics. I can imagine two reasons why they made this choice. The first, and obvious, reason is that they want to communicate with non-mathematical readers. But I suspect that there may have been a second and more interesting reason: they wanted to convey something which cannot be communicated by mathematics alone. If I claim to understand the behaviour of some system, I mean rather more than that I understand the mathematical description of it. I mean that I can in some way analyse it and play with it in my head, imagining how it would behave in various circumstances. For want of an alternative, this ability can be described as having a 'physical intuition' about the system. This should not be taken to mean that one can have such intuitions only about physical systems, since they are just as relevant in biology. More nearly, what is meant is that one can make use of one's everyday experience of physical objects to visualise how imaginary objects might behave. Such an intuition is normally a complement to mathematical description, rather than an alternative to it. If the mathematical analysis of some system predicts that it will behave in a particular way, one usually tries to gain some insight into why it should do so. Personally, if I cannot gain such an insight, I check the algebra, or the computer programme, very carefully, and expect to find a mistake. Scientists seem to differ a good deal in the extent to which they rely on physical intuition as opposed to algebra, but it would be fatal to try to do without either. One reason Eigen and Winkler avoid mathematics may be that they want to encourage their readers to develop their physical intuition.

The method they adopt, in the place of mathematical description, is to draw analogies between physical processes and games. A game has rules but the course of any particular game is also influenced by chance events which, depending on the game, can include variations in the initial position, the throwing of dice or shuffling of cards, and the unpredictable decisions of the players. Real sequences of events resemble games in that they are constrained by laws (e.g. Newton's laws, the laws of genetics), but the actual sequence is influenced by chance. Thus one cannot predict the course of a particular game of chess, even if the opponents have played before, and still less could one predict the course of a hand at bridge before the deal was made, yet in both cases one can predict that certain rules will be obeyed. Similarly, one cannot predict the future history of any given species, but one knows that these processes will be governed by laws. Of course, we have a complete knowledge of the laws of chess, and only a partial knowledge of the laws of physics or genetics, but that does not invalidate the analogy.

Eigen and Winkler do more than point to the general analogy between games and real events. They invent a number of games which mimic particular biological and physical processes. By playing the games, one should acquire an intuition about the processes of which the games are analogues. The approach can best be illustrated by describing the simplest game in the book. This game shows how a system can approach a stable equilibrium by a series of chance events. The game can be played on a board with 36 squares (six by six). Each square is occupied by a bead, which can be black or white. To make a move, two dice are thrown. The numbers on the dice then specify one of the squares on the board – for example, three up and four across. The bead on that square is then replaced by one of the opposite colour. This completes a move. Suppose, for example, that the game is started with all the beads black. After a number of moves, it will be found that the numbers of black and white beads are approximately equal, the numbers fluctuating either side of equality. In this simple form, it is easy to predict what will happen without actually playing the game, although there are questions whose answers are not immediately obvious. For example, what is the distribution of frequencies with which various numbers of black beads are present in a long sequence of throws? How big a board must one have to be reasonably certain that, after reaching the equilibrium, the board will not again come to have beads of only one colour?

This game is an analogue of the way in which equilibria are reached in physical systems. Even relatively simple changes in the rules can lead to quite rich behaviour, which it would be difficult to predict without playing the game. For example, suppose that, once a square has been chosen by throwing the dice, the bead on it is replaced only if some specified minimum number of the beads on neighbouring squares are also of the opposite colour. Such 'cooperative' games can give rise to spatial patterns, in a way analogous to the appearance of patterns in the development of living organisms.

How far do the authors succeed? This is a hard question to answer. When writing about science for non-scientists, it is as well to have an imaginary reader in mind. I have two imaginary readers. One is an intelligent but ignorant 16-year-old: myself when young. The other is an intelligent but even more ignorant British civil servant, bent on improving his mind. How would these two fare with *The Laws of the Game*? The civil servant would, I suspect, fare very badly, if only because it might not occur to him that it is necessary actually to play through the various games in order to gain an insight into the behaviour which follows from particular rules. If this was obvious, there would be no need to acquire physical intuition: we would all be born with it, and Eigen and Winkler are wasting their time. In any case, for all I know they do not allow dicing at the Reform Club.

I fear that my 16-year-old might also fare badly, although for different reasons. He might enjoy playing the games, and if so would gain some insight into physics and biology. The trouble is that there are a great many things in the book he would not understand. There are a fair number of things I found difficult, and I have been thinking about these problems for some time. It is hard to tell how a relatively ignorant reader would react. The right policy would be to press on regardless, accepting that some things will remain obscure, and enjoying the bits that make sense. Some readers, however, may merely get discouraged.

In effect, my fear is that Eigen and Winkler may have attempted the impossible. Eigen himself did brilliant work in chemical kinetics, and then turned to biology. Unlike some other physical scientists who have made this switch, he has grasped the fundamental ideas of biology. In particular, he understands the principle of natural selection, which usually seems to defeat physicists. He has spent some extremely fruitful years working on the origin of life: that is, on how chemical processes gave rise to biological ones.

As a consquence of this history, Eigen has thought hard and
long about the problems discussed in this book. It may be
that the understanding which he and Dr Winkler are trying
to convey can be acquired only by years of work with
science. Perhaps the only way to acquire a well-furnished
mind is to spend a lifetime, or at least a long apprenticeship,
in furnishing it.

For those who would write popular science, this is a
counsel of despair. The authors might offer two replies. The
first is that it is better to convey a vague idea than no idea
at all. They quote with justified disapproval Wittgenstein's
dictum: 'Whatever can be said at all can be said clearly, and
whatever cannot be said clearly should not be said at all.' If
accepted, this would rule out all poetry and much of science,
although it would leave us with most of mathematics and
the London telephone directory.

The second reply they might make is that I have
misidentified their imaginary reader. Perhaps they were
writing for readers with more knowledge of science than I
have supposed: not for me when young but me now. Indeed,
I did find their games illuminating, partly because, being
originally physical chemists, they tend to draw their analogies
in the opposite direction to me. They understand predator-
prey systems by seeing that they are just like the law of
chemical mass action, whereas I understand chemistry by
imagining animals running about. The ideal readers for this
book may be teachers of science. There is a rich source of
ideas here for anyone trying to teach the structure of scientific
thinking. There seems little doubt that young mammals play
games because in that way they acquire skills they will need
later. Why, then, should we not play games to learn science?

25

Matchsticks, Brains and Curtain Rings

The history of biology suggests that we understand how animals do something only when we invent machines that do the same thing. Animal flight was a mystery until engineers understood the aerofoil. To give a more controversial example, the nature of heredity and the genetic code was understood only after the invention of gramophones, tape recorders and other information-processing machines. It is therefore natural to hope that the invention of computers, which do many of the things that brains do, will help us to understand how we think. Is this hope justified?

The one great advantage of programming a computer to simulate a natural process is that you have to say exactly what you mean. If someone programs a computer so that, when shown a picture, it can correctly say 'that is a pyramid standing on a cube', we can be sure that the process that has been programmed does actually work. That is important, because biology is full of verbal assertions that some mechanism will generate some result, when in fact it won't. However, can we be sure that the way the computer does it is in any way similar to the way you and I do it?

Some differences between brains and computers are unimportant. It need not matter that computers are made of transistors and brains of neurones, provided that the way they perform calculations are logically similar. This idea of similarity – more precisely, 'isomorphism' – is so important that I will give an example. Cricket umpires are said to count the balls in an over by moving six pebbles from one pocket to another. It would be easy to give them an electric counter with six switches, arranged so that a bell rang when all the switches were on. If an umpire counted balls by moving a switch after each ball, he would be behaving isomorphically with a pebble-moving umpire. Notice, however, that if he simply said one, two, three and so on as ball' was bowled, this would not be isomorphic, because each sound is different from all the others, and it matters in what order they are made, whereas the pebbles can be all alike, and can be moved in any order.

A deeper difference between computers and brains is that the former are digital devices and the latter analog ones. Unfortunately, the word 'analog' has two distinct meanings, one of which is usually forgotten. I first met an 'analog computer' during the war, when I was working on aircraft design. It was important, early in the design, to estimate how the aircraft would vibrate. We did this by building an electric circuit which would behave in a way analogous to the aeroplane. This can be done because the behaviour of a tuning circuit (which in a radio can be altered to resonate in tune with particular radio waves) is exactly analogous to that of a weight on a spring (both obey the differential equation $d^2x/dt^2 = -kx$). The oscillations of the circuit then predicted how the structure would vibrate.

The old analog computer differed from a modern digital one in two ways. First, electric currents in the analog machine varied continuously, whereas in a digital one each transistor is in one of two states. Consequently, 'analog' has come to mean 'varying continuously'. Now, I think a neurone is best thought of as an analog device. It is true that a nerve fibre either fires or it doesn't: in that sense it is digital. But the frequency of impulses in a fibre can vary continuously, from zero to some upper limit. However, I don't think this really matters. As anyone who has played *Space Invaders* knows, it is easy to get a digital device to behave in a nearly continuous manner. It is in fact possible to build out of transistors, or other all-or-none devices, units which behave very much like neurones.

There is, however, another meaning of the word 'analog' which may matter more. The electrical analog of a mechanical system works because the laws of physics are as they are: if the tension in a spring increased as the square of its extension, the thing wouldn't work. In other words, an analog computer works by simulating one physical system by another one governed by the same mathematical equations. This can sometimes offer a simple way of solving a problem which otherwise would be very laborious. As an example, consider the following classic problem. You want to go by air from London to Moscow in the shortest possible time. You are given a list of 40 cities, including London and Moscow, together with the time taken to fly direct between any pair given that a direct flight exists. Assuming that waiting time is zero, you must find the route for which the sum of the times is a minimum. On a digital computer, the problem is in one sense trivial. You search all possible routes, and choose the shortest. The only snag is that the number of possible routes is immense, so the calculation would take too long to do. There is probably a better way of doing it, by cutting down the number of possible routes that need to be measured: I don't know if one has been found.

The problem is easily solved, however, on a purpose-built analog computer. This device consists of 40 curtain rings, each representing a city. If there is a flight between two cities, the rings are connected by a string whose length in inches equals the flight time in minutes. To find the quickest route, take hold of the London and Moscow rings, and pull them apart: when a set of strings becomes taut, that is the best route.

The point of this example is to show that a physical analog (time corresponds to length: times and lengths can be added by placing them end to end) can sometimes solve a problem in a way which is *not* isomorphic with the way one would program a computer to do it. Could this be true of brains? I think it may be. Consider, for example, the problem of judging distances. This may arise in at least two contexts.

First, how do we judge the apparent distance between two points in our visual field? (I am not asking the harder question of how we judge real distances, allowing for foreshortening.) A topological representation of our visual field is present in the visual cortex of the brain, with particular points on the retina corresponding to points in the cortex. Hence the problem reduces to that of how one estimates the distance between two neurones in one's brain.

The second context is that of judging the distance between two places we have visited. It seems that humans, and some animals, do form 'cognitive maps': that is, they have some representation in their heads which carries the same information that a map would convey. To digress for a moment, it is not easy to design experiments that will distinguish between an animal which finds its way from A to B by using such a map, and one that does so by following some learnt rule of thumb, such as 'first left, second right'. However, that would take me too far from the topic of this article, so I shall assume that at least some animals do form maps. It does not follow that the representation consists of a sheet of tissue in which particular neurones correspond to particular places, and in which the neurone map is topologically similar to the real one, but this seems much the most plausible assumption. If so, the problem of judging distances also reduces to measuring the distance between two neurones.

Before asking how the brain might measure such a distance, I want to describe two ways of measuring distance which are *not* isomorphic to one another. I give you a one inch to the mile map, and a box of matches, and ask you to find the distance from London to Brighton. One way would be to take map references of the two towns, say (60,10) and (20,40), and use Pythagoras's theorem (which relates the lengths of the sides of a right-angled triangle: $a^2 + b^2 = c^2$, where c is the longest side of the triangle) to find the distance: $[(60-20)^2 + (10-40)^2]^{1/2} = 50$ miles. This way is isomorphic to the way one would program a digital computer to do the job. Alternatively, you might notice that the matches are one inch long. If so, you would arrange them end to end in a straight line between the towns, and count the matches.

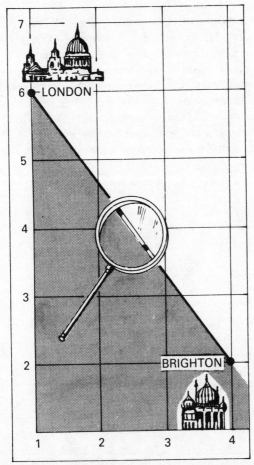

Figure 25.1 Maths or matchsticks: two ways to estimate distance.
Which does the brain use?

I think the latter method is more likely to be isomorphic
with the way the brain makes such estimates. For example,
the distance between neurones A to B might be measured
by the time a message takes to go from A to B and
back again. The alternative, of using Pythagoras, seems
implausible. Indeed, I doubt whether our brains add or
multiply in ways isomorphic to a computer. The latter uses
a place notation, just as an abacus does, with different

elements corresponding to units, tens, hundreds, etc (or, more usually, 2s, 4s, 8s, etc). I would be surprised if there is anything isomorphic to this in our brains.

If I am right about this, it casts some doubt on the belief that, by understanding how computers solve problems, we shall learn how brains do. However, it is unlikely that I am right. I have been doing something that I disapprove of strongly when it is indulged in by others: I have been pontificating in public about matters of which I am largely ignorant.

26

Hypercycles and the Origin of Life

Perhaps the most difficult step to explain in the origin of life is that from the replication of molecules (RNA for example) in the absence of specific proteins, to the appearance of polymerases and other proteins involved in the replication of RNA and themselves coded for by that RNA. Suppose we start with a population of replicating RNA molecules. Without specific enzymes the accuracy of replication is low and hence the length of RNA which could be precisely replicated small. Before replication can be reasonably accurate, there must as a minimum be a specific polymerase, as well as synthetases and tRNAs, which in turn implies an RNA genome of considerable length. Thus, even if one supposes an initially very limited set of codons, one cannot have accurate replication without a length of RNA, say, 2000 or more base pairs, and one cannot have that much RNA without accurate replication. This is the central problem discussed in a series of papers by Manfred Eigen and Peter Schuster proposing the 'hypercycle' as a necessary intermediate stage.

First, imagine a population of replicating RNA molecules, lacking genetic recombination, but with a certain 'error' or 'mutation' rate per base replication. Very roughly, if more

than one mutation occurred per molecular replication, the
population would come to consist of a random collection of
sequences. But if less than one mutation occurred per
replication, and if one sequence was 'fitter' than others in
the sense of being more stable and/or more easily replicated,
then a population would arise whose sequences were centred
around this optimal one. Such a population Eigen and
Schuster call a molecular 'quasi-species'. If the mutation rate
per replication was much less than one, and if fitness
differences were large, most molecules would have the
optimal sequence. As the mutation rate increased, the
proportion of the population with the optimal sequence
would fall, until a critical mutation rate was reached, above
which sequences would be random.

The replication of RNA molecules in the absence of a
specific replicase has not yet been achieved in the laboratory.
It may initially have depended on the presence of non-
specific random-sequence polypeptides. At this stage, the
error rate would depend solely on the energy levels of base
pairing between G and C, and between A and U, and would
be such that the maximum length of a quasi-species would
be 10–100 base pairs; even this length would require the
RNA to be G–C rich. In contrast, RNA replication by the
enzyme coded for by the RNA phage Qβ is accurate enough
to permit a genome of 1000–10,000 base pairs, which would
be sufficient to code for a primitive protein-synthesising
machinery. The next stage in the evolution of accuracy of
replication, achieved by prokaryotes, was the recognition
and correction of mismatches in DNA replication, a process
which depends on recognising which is the old strand and
which the new. With such proof-correcting, a genome of at
least 10^6 base pairs, and perhaps 10^8–10^9 base pairs, can be
maintained. It is the gap between the 10–100 bases, without
specific enzymes, and the 1000–10,000 bases, with specific
enzymes, that Eigen and Schuster aim to bridge.

To understand their proposal, imagine that you wish to
replicate the message GOD SAVE THE QUEEN. Counting
5 bits per letter, this is a total of 75 bits, and a maximum
of 25 bits per word. Supposed the error rate was, say, 1/50
per bit. You start with a population of messages, copy each

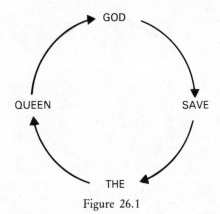

Figure 26.1

of them with that error rate, and then discard half of the population, applying a rule of selection so that messages with many errors are more likely to be discarded. Despite this 'natural selection', the population would steadily accumulate errors. The mutation rate is too high.

Suppose, therefore, that after each replication you selected word by word, discarding half the quasi-species GOD, and similarly for each other word. Since no word contains as many as 50 bits, you could now maintain a meaningful message. However, there is an important reason why this is *not* an adequate model of the replication of a set of RNA molecules. Thus, to maintain complete messages you would have to ensure in each generation that you selected equal numbers (or approximately so) of each word. But suppose the words competed, as RNA molecules would compete for substrates. Then before long only one quasi-species would remain – GOD only, or THE only, if short words replicate faster.

Hence if the words are strung together into a single selected message (analogous to a single RNA molecule), the message is too long to replicate; if they are replicated separately, competition between the words destroys the message. You can escape from this dilemma by arranging the words in a 'hypercycle', as in Figure 26.1. In this cycle, the *rate* (not the accuracy) of replication of SAVE is increased by the number of GODs, of THE by the number of SAVEs,

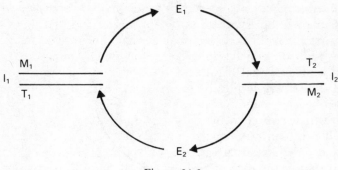

Figure 26.2

of QUEEN by the number of THEs, and, coming full circle, of GOD by the number of QUEENs. Each word, or quasi-species, is selected independently, according to its own fitness. This is a hypercycle. It can be shown that only by linking together a set of replicating quasi-species in this cyclical way can the stability of the whole be maintained.

A possible way in which a simple 'two-word' hypercycle might be realised in molecular terms is shown in Figure 26.2. (Eigen and Schuster would not insist on this particular realisation; it is intended only to show the kind of thing they have in mind.) It is necessary to describe a cycle in some detail, so as to be able to discuss the difficulty which arises in explaining the further evolution of such cycles. I_1 and I_2 are two RNA quasi-series. Both the + and − strands must be good replicators. RNA molecules have a phenotype, because they fold up, and folding patterns affect both survival (resistance to hydrolysis) and replication rate. Since both + and − strains must replicate, they are likely to have similar folding patterns. Further, I_1 and I_2 may be descendants of the same ancestor quasi-species, in which case they will resemble one another.

In this particular realisation, the − strands T_1 and T_2 are supposed to be 'adapters' or precursors of tRNA, coupling with specific amino acids, and having an anticodon loop. The + strands M_1 and M_2 are 'messengers', composed of only two kinds of codon, and coding for two proteins E_1 and E_2 which act as replicases for I_2 and I_1 respectively.

Thus the model assumes that some kind of translation is possible without synthetases or ribosomes, and that the code is arbitrary from the start (that is, that there is no chemical constraint on which triplet codes for which amino acid). It is not an essential feature of the argument that messenger and transfer molecules be the + and − strands of the same quasi-species, but if they were not, then a hypercycle of more than two elements would be needed.

One difficulty in conceiving how such a cycle might arise is that it requires that each of two 'messengers' (or of more than two in the case of a longer cycle) should programme a replicase for the other: Eigen and Schuster argue that this becomes less implausible if it is remembered that I_1 and I_2 may be descended from members of the same molecular quasi-species. Initially, the members of a quasi-species would be sufficiently alike to share the same replicase. Hypercyclic organisation could then arise by the gradual differentiation of a single quasi-species into two or more.

I now digress to give the authors' views on the origin of the code. They argue that primitive RNA molecules must have been G–C rich (for reasons already mentioned). Further, in the absence of ribosomes the message must have been one which could only be read 'in frame'. These two conditions imply that the earliest messages consisted of strings either of GNC codons or CNG codons, where N stands for any base. The authors use an argument based on wobble to prefer the former codon type. Hence, from arguments based on the stability, replication and translation of nucleic acids, they conclude that the first two codons were GGC and GCC, followed by GAC and GUC. This conclusion agrees nicely with the likely abundances of amino acids in the primitive oceans. By far the most abundant amino acids in simulated prebiotic synthesis are glycine and alanine (today coded by GGC and GCC), the next commonest being aspartic acid (GAC) and valine (GUC).

Returning to the main theme, the following question arises. Given that a hypercycle ensures the accurate replication of a larger total quantity of information, how will it evolve further? Consider the particular realisation in Figure 21.2.

There are three kinds of mutation which might occur in I_1 (and a similar set in I_2):

(1) Mutations making I_1 better (or worse) at replicating, for example by becoming a better target for E_2. (I_1 might also become a better target for E_1; if this process went too far the hypercycle would be disrupted.)

(2) Mutations making E_1 a better (or worse) replicator of I_2.

(3) Mutations making T_1 a better adapter.

If there is not compartmentalisation, only mutations of type (1) would be incorporated by selection. Each quasi-species in the hypercycle would evolve independently. There would be no selection favouring mutations of types (2) and (3), although such mutations would be needed before the speed and accuracy of replication would improve sufficiently to permit the genetic information to be united in a single 'chromosome'. There is a natural analogy here to an ecosystem. Imagine an ecosystem consisting of grass, antelopes and lions. The more grass there is, the more rapid is the multiplication of antelopes. Similarly, antelopes encourage the multiplication of lions, and (by some stretch of the imagination) lions, by fertilising the ground, encourage the growth of grass. The point of the analogy is this. Natural selection will favour mutations in grass which increase the fitness of individual grass plants, but will not favour changes in grass making it more edible to antelopes.

How then can a hypercycle evolve characteristics which favour the growth of the cycle as a whole, rather than merely its constituent parts? So long as there is no compartmentalisation, it cannot. For natural selection to act, there must be individuals. In the present context, this seems to require that the compartments of a hypercycle be enclosed in a membrane to form a proto-cell. If these proto-cells grew and divided – by some kind of budding or fission – at a rate which increased with the growth rate of the enclosed hypercycle, then selection would favour mutations making the hypercycle more efficient. In the example of Figure 26.2, mutations of types (2) and (3) would be favoured.

Clearly, these papers raise more problems than they solve. Their merit is that they put in sharp terms the problem raised by the relatively inaccurate replication of nucleic acids, and, in the hypercycle, they suggest a way in which a relatively complex structure could be replicated despite this inaccuracy.

27

Popper's World

Karl Popper is perhaps the only living philosopher of science who has had a substantial influence on the way scientists do what they do. I say 'perhaps' because the same claim might be made for Thomas Kuhn. However, Kuhn seems to me a perceptive sociologist of science, but a poor philosopher. Also, in so far as he has had an effect on the way scientists behave, it has been pernicious: to be a great scientist, according to Kuhn, you must do revolutionary science, and the best evidence that you are doing it is that you are so obscure and inconsistent in your statements as to be wholly incomprehensible to others. This does not make for good science. In contrast, Popper has encouraged us to speculate boldly, but to be fiercely critical; it is true that we usually manage the latter only for other people's ideas, and not for our own, but since science is a social activity, that suffices.

Popper's main contribution to the philosophy of science was contained in his *Logic of Scientific Discovery*, first published in 1934, and in English translation in 1959. It was during the process of translation that Popper wrote a mass of additional material, which was at first intended to appear as an appendix to the *Logic*. However, the material accumulated until it was longer than the original book. Most

of it is now published for the first time, in the form which
it had reached by 1962. It appears in three volumes: *Realism
and the Aim of Science*; *The Open Universe: An Argument
for Indeterminism*; and *Quantum Theory and the Schism in
Physics*. This review treats only the second of these volumes,
because that is as far as my competence will stretch. Indeed,
many would doubt my competence to review even this
volume, since I lack any training in philosophy. My excuse
is that I have spent much of my life thinking about a branch
of science – evolutionary biology – to which questions of
chance and necessity are central. Also, although I regret not
tackling the other two volumes, I think this is the one of
most interest to the general reader.

As is made clear in his autobiography *Unended Quest*,
Popper's life has been dominated by two problems, both
presented to him during his youth in pre-Nazi Vienna. The
first was to diagnose the difference between what seemed to
him to be genuine knowledge, as represented by the work
of Einstein, and what he saw as mere rhetoric, as represented
by Freud, Marx, Adler and Jung. It was this problem which
led him to the famous demarcation criterion of falsifiability,
and to *The Logic of Scientific Discovery*. The second problem
was to find a philosophical justification for human freedom;
it is this which underlies the *Postscript*.

That this is Popper's preoccupation is made clear in the
Preface to this volume, in which he writes: 'This book is
then a kind of prolegomenon to the question of human
freedom and creativity.' He points out that there is an
apparent contradiction between the widely held view that
every event has a cause, and the common-sense conviction
that, at least sometimes, men are free to choose what they
will do – that men are sometimes masters of their fate. It is
this apparent contradiction which he sets out to resolve.

The strongest version of the determinist thesis is due to
Laplace; the first part of this review will therefore be
concerned with Popper's refutation of Laplace's argument.
I then digress briefly to discuss Popper's 'propensity' theory
of probability, which, although not essential to the main
theme of the book, is of real interest to scientists. Finally,
I discuss Popper's more recent views on the freedom of

human action. At this point, reluctantly, I find myself in total disagreement with him. I suspect that this is because my view of the world is conditioned primarily by biology, and his by physics.

Laplace, the great formaliser of Newtonian physics, argues as follows. If some universal intelligence could know all the laws of physics, and the present state of the universe (in Newtonian terms, the positions and velocities of all particles in the universe), then it could calculate the future with complete certainty. In a sense, therefore, the future already exists. To use a modern analogy, the universe – past, present and future – resembles a cine film: you and I are restricted to observing only the contemporary frames, but the future frames exist, with just the same reality as the past. Nothing we can do will alter that future: free will is an illusion.

I remember being totally persuaded by this argument when I was about sixteen, although I do not think it made me any less wilful. More recently, I have rejected it on the grounds that the knowledge, and the calculations, are in fact impossible. No calculator smaller than the universe itself could contain the necessary information. I suppose that it is true that the universe knows what it is going to do, in the sense that *something* is going to happen, but nothing smaller than the universe can possibly know what.

Popper offers similar but more coherent grounds for rejecting Laplace's argument. He says that he does not regard philosophical determinism – the cine-film view of the universe – as ultimately refutable. He is concerned only to show that it does not follow from our knowledge of science. Further, it would not follow if we took an entirely Newtonian view of physics: Popper's rejection of determinacy does not rest on the indeterminacy of quantum physics, which he is in any case unwilling to accept at face value – but that is an argument I do not want to get into. He argues that it is quite possible to hold that every event has a cause, and yet to reject Laplace's argument.

The accuracy of one's predictions, Popper points out, depends on the accuracy with which one knows the present. Since one can never know the present with absolute precision, Laplacean determinism requires that one answer the following

question: 'Given that I want to know the future with some
specified degree of 'accuracy, how accurately must I know
the present?' Quite rightly, Popper argues that this question
is unanswerable, because a vanishingly small difference in
initial conditions can give rise to an indefinitely large
difference in end-results. He quotes a mathematical result of
Hadamard's (of which I was unaware) to prove this. Today
he would probably refer to the mathematics of 'chaos',
which explains such apparently random and unpredictable
phenomena as turbulence in terms of a fully deterministic
model. Popper gives other reasons for rejecting the Laplacean
argument, including the reasonable one that no calculator
can predict its own future with certainty.

I now digress to discuss Popper's propensity theory of
probability. When I first met this (in one of the actually
published appendices to *The Logic of Scientific Discovery*),
I found it hard to believe that a man of such clarity of mind
could hold such an odd view. Rereading it after twenty
years, I found myself accepting it like an old friend. I can
best explain Popper's theory by first describing what might
be called the old-fashioned scientist's view of the matter. If
you toss a penny, it comes down heads about half of the
time (unless you happen to be Rosencrantz, or was it
Guildenstern?), but any particular throw is unpredictable.
However, if you measured exactly the initial position,
velocity and spin of the penny, you could calculate exactly
what would happen; the initial conditions determine the
outcome, in a thoroughly Laplacean manner. To this, Popper
would reply: how do you account for the fact that the initial
conditions, on different tosses, are such as to generate a
random sequence, with probability one-half? Indeed, he
might trap the old-fashioned scientist in an infinite regress:
the initial conditions were as they were because *their* initial
conditions were as they were, and so on *ad infinitum*. Is it
not better simply to say that a penny has a propensity of
one-half to give a head? This propensity is a property of the
penny, just like its mass or its colour. As I say, this seemed
idiotic when I first met it. Now it seems to fit rather well
with the attitude I have come to accept (originally acquired,
I think, from J. B. S. Haldane): that a random event is an

event into whose causes it is not at present efficient to inquire. Popper would perhaps replace 'at present' by 'ever'.

So far, so good. But now I come to the parting of the ways. What is the relation between philosophical determinism and the concept of human freedom? For Popper, the whole motive for wishing to dethrone determinism is to underwrite freedom. Now of course there is free will. I am writing this review of my own free will: no one made me do it. But does this mean that my action was uncaused? A psychologist might say that it was caused by my respect for Popper, by my wish to dissociate myself from his views on free will, or even by my knowledge of the state of my bank balance. But my 'respect', my 'wish' and my 'knowledge' are, or so I would hold, represented by physical states of my brain. If not by physical states of my brain, then how do they come to influence the movements of my hand as I write?

I think that Popper would agree that a free act, such as the writing of this review, can be caused. However, in collaboration with Eccles, he has developed a way of talking about causation which seems to me mistaken. I think it is relevant to discuss these ideas here, because they are treated in addenda to the present volume, written since 1962, and also because they form a continuation of Popper's search for a philosophical justification for freedom. He proposes that we should recognise three 'Worlds'. World 1 is the world of things – of pens, paper, neurones and muscles. World 2 is the world of consciousness. World 3 is the world of cultural artifacts, in their conceptual rather than their physical form. The concept of falsification is part of World 3, but the many copies of the book in which it was proposed are each part of World 1. This is fine. Popper then goes on to say that World 1 is 'causally open' to World 2, and that World 2 is causally open to World 3. This is also acceptable, provided we are careful. The concept of falsification (World 3) may influence the way I think about an experiment (World 2), and this in turn may influence which instruments are connected in what ways (World 1). But in that last sentence, the 'concept of falsification' which influences what I do is not an immaterial concept, but the physical representation of that concept in my brain, and so is 'the way I think about an experiment'.

Popper and Eccles appear to think that concepts and feelings are causes of actions, independent of the events in the brain which embody those concepts and feelings. Now just as Popper admits that philosophical determinism (the cine-film view of the universe) is irrefutable, I admit that I cannot refute the idea that a feeling can be an immaterial cause of a material event. We must be clear what this means. As I write this review, my hand moves, because the muscles in my hand and arm contract, because neurones descending from my brain and spinal cord carry impulses which cause them to contract. The Popper–Eccles view, if I understand it (and if I do not, the fault is not entirely mine), is that at some point this chain of material causation is broken, and something immaterial causes a neurone to fire.

There is no way of showing that this is not so. Why, then, do I object to supposing that it is? Fundamentally, because I would like to know how brains work, and if one once admits that the interesting things that brains do are not caused by material events, the question becomes unanswerable. I object because I think Popper's views are defeatist. I suggested earlier that my disagreement with Popper arises because I am a biologist. I am not altered in this opinion by the fact that Eccles is a distinguished neurobiologist, because it is an occupational risk of biologists to claim, towards the end of their careers, that problems which they have not solved are insoluble. But I think that Popper is sometimes too ready to treat as insoluble problems I would like to see solved. This is most vividly illustrated by his remark: 'It seems reasonable to regard the emergence of consciousness and previously that of life as two comparatively recent events in the evolution of the universe; as events which, like the beginning of the universe, are at present, and perhaps for ever, beyond our scientific understanding.' I agree that we are not making much headway with consciousness right now, but I hope to live to see the solution to the problem of the origin of life.

28

Rottenness is All

This is an ambitious book which suggests that a new picture of the nature of the universe is emerging from the study of thermodynamics, and that this picture will heal the breach between the scientific and the poetic view of man. Prigogine's distinction as a scientist – he won the Nobel Prize for Chemistry in 1977 – requires that we take his views seriously, at least on the first of these claims. The major part of the book is devoted to explaining, in non-mathematical language, the new science that the authors see emerging, and to which Prigogine helped to give birth. The ideas are hard, but I think they succeed. There are places, particularly in their treatment of quantum theory, where readers without some previous knowledge may lose the thread; certainly I did. But anyone prepared to make a serious effort will get some insight into what is happening.

In this review, I aim to do three things. First, I shall explain the new thermodynamics; here I shall try to follow the authors' account, not criticise it. Second, I will say something about the relevance of their ideas to biology. Finally, I shall comment briefly on their significance for man's view of himself.

The authors see the history of the physical sciences as

dominated by two apparently incompatible approaches, dynamics and thermodynamics. Dynamics, stemming from Galileo and Newton, sees the world as deterministic and reversible. As Laplace argued, it follows from the dynamical world picture that, given the positions and velocities of all the particles in the universe at any instant, an all-powerful intelligence could calculate the future. It also follows that if, at some instant, the direction of motion of every particle was exactly reversed, then the universe would return to the states it had occupied at earlier times. To put the same point in another way, if a film of some sequence of events was run forwards and backwards, it would be impossible for a viewer to tell which direction was correct.

This deterministic and reversible picture is as true of Einstein's dynamics as of Newton's. It has no place for our subjective view of time. It is as if the whole past and future were eternally present, although we are forced to experience events in a particular sequence, just as, in the cinema, the whole reel of film is present all the time, but we are forced to view the frames in sequence. This dynamic picture seems to be contradicted by everyday experience, not only of subjective time, but also of the irreversible nature of most events. Shown a film, we have no difficulty in deciding whether it is being run in the correct order. Smoke does not flow towards a cigarette end as the cigarette lengthens.

Since the beginning of the nineteenth century, scientists have been increasingly interested in processes that do not run backwards. This interest originated from an effort to understand heat engines. A steam engine converts the chemical energy in coal into mechanical work: you cannot run it backwards, to turn mechanical work into coal. It will help to consider a simpler example of an irreversible process. A drop of water falling on the surface of a lake causes waves, which spread out as concentric rings, diminishing in amplitude as they go, until finally they disappear, leaving the lake smooth once more. One never sees a series of concentric rings converging on a point, and propelling a drop of water into the sky. If we ask what has happened to the energy of the water drop when the lake finally becomes smooth, the answer is that the velocity of the water molecules has increased infinitesimally: in other words, the lake has got a little hotter.

Thermodynamics arose as a way of describing and predicting the behaviour of systems in which irreversible changes occur, and in which energy is converted from one form to another – heat, mechanical work, electrical energy, chemical energy, and so on. The formulation of thermodynamics included the famous – I am tempted to say notorious – second law. This law differs from anything in classical dynamics, because it asserts that changes will occur only in one specific direction. It states that a certain quantity, entropy, can only increase (or, at equilibrium, remain constant) in a physically closed system. Increase in entropy really requires a mathematical definition, but can be understood qualitatively as saying that things will get more mixed up, random and unstructured. A vessel containing hot water at one end and cold at the other will come to contain lukewarm water everywhere, but a vessel containing lukewarm water will not separate out into hot and cold. When our drop of water falls into the lake, there is initially a set of molecules all moving in the same direction: finally, the directions are random.

At first sight, thermodynamics seems to have no more room for life than dynamics. Increase of entropy can explain death, but not life: why we rot, but not how we ripen. Indeed, the second law is regularly trotted out by the creationists as a reason why life could not have arisen without miraculous intervention. Before tackling this difficulty, however, I must say something about the attempts that have been made to reconcile reversible dynamics and irreversible thermodynamics. The crucial figures were Maxwell and Boltzmann. We can imagine a gas as consisting of elastic particles colliding with one another. The faster the particles move, the higher the temperature. Suppose that at one end of a container the particles are moving much faster than at the other. Boltzmann showed that the effect of a collision between a fast-moving and a slow-moving particle is, more often than not, to make their velocities more nearly equal. Consequently, the effect of collisions is an averaging out of the velocities: the gas becomes the same temperature everywhere. In fact, the velocities of the molecules do not become identical, but have a distribution around an average value. When an equilibrium is finally reached, this distribution no longer changes with time.

Thus a dynamic model of molecules obeying Newton's laws can predict the increase of entropy demanded by the second law. But there is a snag, contained in the phrase 'more often than not' that I used above. Let us again return to the drop of water falling into a lake. Suppose that, when the waves have disappeared, the direction of motion of every molecule is exactly reversed. Does not the dynamic model predict that history will run backwards, and that converging rings would appear and propel a drop into the sky? Indeed it does. How then does Boltzmann's argument help, since a dynamic model can predict either an increase or a decrease of entropy? The answer is that, for a flat surface to generate converging rings, the initial positions and velocities of every molecule in the lake would have to be precisely specified. If these values were varied, even slightly, no pattern would appear. In contrast, if we start with the falling drop, the initial values could be varied extensively, provided certain averages were maintained, and the visible behaviour would be unaltered.

Hence, from the point of view of dynamics, the law that entropy increases is merely probable. There are initial conditions for which entropy in a closed system will decrease, but they are very unlikely. Prigogine and Stengers argue that, in a technical sense, they are infinitely unlikely: it is this fact which, in their view, guarantees irreversibility, and the arrow of time.

Does this mean that there is some justice in the creationists' claim that the origin of life required a miracle? After all, the entropy of a living organism is less than that of the non-living matter from which it emerged. The short answer to this is that the second law applies only to a closed system: that is, a system physically isolated from everything else. There is, however, a more interesting answer to the question, which emerges from a study of systems which are a long way from thermodynamic equilibrium; it is in this field that Prigogine's work earned him his Nobel Prize. The point can best be explained by an example. If the plug is pulled out of a basin, the escaping water may form a vortex, with the water spinning round a central hole. If water is continuously added, so as to maintain the level in the basin, the vortex will persist. The movements of the water molecules in the vortex are far from random: in fact, they move coherently to form an ordered structure.

Such structures Prigogine has called 'dissipative structures'. Their essential characteristic is that they require a continuing input of energy from without. In the case of the vortex, water must be continuously added to the basin. If that ceases, the basin will empty, and the water will flow to the lowest point it can reach, at which point all structure will disappear: the system has reached equilibrium. If a system is far from equilibrium, however, dissipative structures will form. Two other examples will illustrate the point. First, suppose that a thin layer of water is heated from below. If the temperature difference between top and bottom is great enough, an ordered flow of liquid is set up, in a hexagonal pattern known as Benard cells.

A second example, which has been of particular interest to biologists, is the chemical reaction known as the Zhabotinsky reaction. Chemical reactions proceed until an equilibrium is reached; the substances produced by the reaction become randomly mixed by a process of diffusion, so that, typically, all substances are uniformly distributed in space. In the Zhabotinsky reaction, however, a spatial pattern appears. Since some of the reactants are coloured, bands and rings of colour appear, and move slowly across the region in which the reaction is taking place. The first person to show mathematically that chemical reaction and diffusion can give rise to large-scale spatial patterns was Alan Turing, in 1952, shortly before his death.

In each of these examples – a vortex in a basin, Benard cells, Zhabotinsky's reaction – there is an external supply of energy: the potential energy of the water, the heating from below, the chemical energy of the reactants. In each case, an ordered structure appears. The mathematical analysis of such structures, and of the conditions for their origin and maintenance, constitutes the new approach to science that

Prigogine and Stengers describe. It is clear that living organisms can be viewed as dissipative structures. Life requires a continuous input of energy in the form of food or sunlight. So long as that input continues, complex structures can exist.

It is, therefore, somewhat surprising that few biologists have taken this view of life very seriously, although, as I shall explain, this neglect is not universal. A typical account of a living organism would start with a description of how the chemical reactions going on in cells are controlled by proteins, and go on to explain how the structure of those proteins is coded for in the genetic material, the DNA, and how, in evolutionary time, the DNA has been programmed by natural selection. As the terms 'control', 'code' and 'programme' indicate, the central concepts are derived from cybernetics and information theory. As Dawkins put it in *The Selfish Gene*, organisms can be viewed as 'robot vehicles blindly programmed to preserve the selfish molecules known as genes'. There is not much room here for the organism as a vortex.

Some biologists, however, would argue that to think only of controls and programmes is to forget that there has to be an object to be controlled: 'You use the snaffle and the curb all right, but where's the bloody horse?' To such people, Prigogine's ideas will have an appeal. In the main, they are working on development. Developmental biologists tend to fall into two schools. For some, all that is needed is to understand how the information in the DNA is translated into adult structure; they seek to discover how groups of genes are switched on and off in different tissues. For others, the fundamental problem is the appearance, during development, of a spatially complex structure from a relatively homogeneous egg. Zhabotinsky's reaction seems a better model than a computer programme for this process. Of course, no one imagines that Zhabotinsky's reaction is more than an analogy, but it does seem to be the *kind* of process we should be looking for. A number of biologically more plausible processes have been proposed, and are being investigated. Ultimately, of course, there need be no contradiction between these two approaches. Even if we come to view embryos as dissipative structures, genes can still control those structures by determining which enzymes shall be present, and hence the rates at which chemical reactions shall proceed. But for the present, there is heated disagreement about what is the most fruitful way forward.

As it happens, I do not think that the biologists engaged in this enterprise have been directly influenced by non-equilibrium thermodynamics. Certainly my own interest was triggered in the Fifties by Turing's paper. It is nice to know that the processes we are thinking about do not contradict the laws of thermodynamics, but in a sense we knew that already, or rather, we knew that if there was a contradiction, it would be so much the worse for thermodynamics. After all, eggs do turn into adults. What I cannot tell is whether the new thermodynamics is going to be of any more detailed use in analysing development. At present I confess I do not see how.

If I am uncertain about the relevance of non-equilibrium thermodynamics for biology, I am still more so about the claims the authors make for its poetic significance. Prigogine has remarked elsewhere: 'I believe it is more helpful, more exhilarating, to be embedded in a living world than to be alone in a dead universe. And this is really what I try to express in my work.' I can understand his discomfort with Einstein's universe. Who would wish to live in an eternal and unchanging four-dimensional universe, and be forced, for no reason that could be understood, to move irrevocably along one of the dimensions, time? Better, surely, to be a changing and developing structure, of a kind that is both natural and predictable in a universe far from thermodynamic equilibrium. I agree, but only because the latter view is better science. A physics which does not permit the occurrence of birth, life and death (in that order) is bad physics.

I do not think we should embrace scientific theories because they are more hopeful, or more exhilarating. I would like to be able to say that we should embrace them because they are true, but that we can never know. The best we can do is to embrace them because they explain a lot of things, are not obviously false, and suggest some interesting questions. I feel sensitive on this matter because, as an evolutionary biologist, I know that people who adopt theories because they are hopeful finish up embracing Lamarckism, which is false, although perhaps not obviously so, or Creationism, which explains nothing, and suggests no questions at all. If non-equilibrium thermodynamics makes poets happier, so be it. But we must accept or reject it on other grounds.

Acknowledgements

The author and the publishers wish to thank the following publishers for kind permission to reproduce the following articles in this book.

Chapter 1, How to win the Nobel Prize: first published as a review of James Watson's *The Double Helix*, in *The Listener*, Vol. 79, No. 2042, 16 May 1968.

Chapter 2, Storming the Fortress, first published as a review of Ernst Mayr's *The Growth of Biological Thought* in *The New York Review of Books*, May 1982.

Chapter 3, Symbolism and Chance: first published in *Scientific Philosophy Today*, ed. J. Agassi and R. S. Cohen, D. Reidel Publishing Co., 1981.

Chapter 4, Science and the Media: first published as Presidential Address to the Zoological Section of the Association for the Advancement of Science, 1983.

Chapter 5, Molecules are not Enough: first published as a review of Richard Levins and R. C. Lewontin's *The Dialectical Biologist*, in *The London Review of Books*, February 1986.

Chapter 7, The Birth of Sociobiology: first published in *The New Scientist*, September 1985.

Chapter 8, Models of Cultural and Genetic Change: first published as a reivew of L. L. Cavalli-Sforza and M. W. Feldman's *Cultural Transmission and Evolution*, and C. J. Lumsden and E. O. Wilson's *Mind and Culture*, in *Evolution*, 1982.

Chapter 9, Constraints on Human Nature: first published as a review of E. O. Wilson's *On Human Nature*, in *Nature*, November 1978.

Chapter 10, Biology and the Behaviour of Man: first published as a review of Philip Kitcher's *Vaulting Ambition*, in *Nature*, November 1985.

Chapter 11, Tinkering: first published as a reivew of Stephen Gould's *The Panda's Thumb*, in *The London Review of Books*, September 1981.

Chapter 12, Boy or Girl: first published as a review of Eric Charnov's *The Theory of Sex Allocation*, in *The London Review of Books*, February 1985.

Chapter 13, Genes and Memes: first published as a review of Richard Dawkins' *The Extended Phenotype*, in *The London Review of Books*, February 1982.

Chapter 14, Natural Selection of Culture?: first published as a reivew of R. Boyd and D. Richerson's *Culture and the Evolutionary Process*, in *The New York Review of Books*, November 1986.

Chapter 15, Palaeontology at the High Table: first published in *Nature*, May 1984.

Chapter 16, Current Controversies in Evolutionary Biology: first published as chapter 10 of *Dimensions of Darwinism*, ed. M. Grene, Cambridge University Press, 1983.

Chapter 17, Did Darwin Get it Right?: first published in *The London Review of Books*, June/July 1981.

Chapter 18, Do we Need a New Evolutionary Paradigm?:

first published as a review of M. W. Ho and P. T. Saunders' (eds) *Beyond Neo-Darwinism*, in *The New Scientist*, 14th March 1985.

Chapter 20, The Limitations of Evolutionary Theory: first published in *The Encyclopaedia of Ignorance*, ed. R. Duncan and M. Weston-Smith, Pergamon, 1977.

Chapter 21, The Evolution of Animal Intelligence: first published in *Minds, Machines and Evolution*, ed. C. Hookway, Cambridge University Press, 1984.

Chapter 22, Evolution and the Theory of Games: first published in *American Scientist*, Vol. 64, No. 1, January/February 1976.

Chapter 23, The Counting Problem: first published in *Towards a Theoretical Biology*, ed. C. H. Waddington, Edinburgh University Press, 1986.

Chapter 24, Understanding Science: first published as a review of Manfred Eigen and Ruthild Winkler's *The Laws of the Game*, in *The London Review of Books*, June 1982.

Chapter 25, Matchsticks, Brains and Curtain Rings: first published in *New Scientist*, February 1984.

Chapter 26, Hypercycles and the Origin of Life: first published in *Nature*, August 1979.

Chapter 27, Popper's World: first published as a review of Karl Popper's *The Open Universe*, in *The London Review of Books*, August 1983.

Chapter 28, Rottenness is All: first published as a review of Ilya Progogine and Isabelle Stengers' *Order out of Chaos*, in *The London Review of Books*, March 1984.

Index

Adaptationist Programme, 87–8
Alexander, R., 46, 89, 91
Analog Computers, 232–3
Aquinas, 9, 12
Aristotle, 8, 12
Attenborough, D., 26
Axelrod, R., 197

Baboons, 198, 212–14
Bateson, W., 12
Bauplans, 154–6
Bean-bag Genetics, 13–14
Bellamy, D., 26
Bernal, J.D., 44–5
Bernal Lecture, 2
Boltzmann, L., 252–3
Boyd, R., 114–21, 259
Buffon, G.L., 9
Burke, James, 29

Cain, A.J., 135
Calculus, 6
Canalization, 127
Cardano, J., 20
Cavalli-Sforza, L.L., 62–7, 76–7, 117, 259
Chagnon, N., 70, 91
Chargaff, E., 3
Charlesworth, B., 142
Charnov, E., 98–104, 259
Chetverikov, S.S., 13
Chinese Science, 17–18
Chomsky, N., 58
Cladism, 133–4, 158
Creationism, 25

Crick, F.H.C., 3, 5, 6, 48
Cuvier, G., 12

Darlington, C.D., 110, 115
Darwin, C., 9, 10, 13, 20, 23, 24, 27, 39, 43–6, 94–5, 114, 127, 152–3, 156–57, 165, 175
Dawkins, R., 28, 35, 55, 59, 77, 105–13, 255, 259
Demographic transition, 65
Determinism, 245–8, 251
Developmental Constraints, 142–3, 151–2, 159–60
de Vries, H., 11
Dialectics, 34–7
Dickemann, M., 91
Dissipative Structures, 254
DNA, 4–7, 34, 49, 60, 107–9, 167, 184, 189, 255
  transfer in prokaryotes, 172
Dodzhansky, Th., 8
Dorze, 15–16, 21, 40–2
Double Helix, 3–7
Dover, G., 128
Drift, genetic 188
Drosophila, 175–6

Eccles, J.C., 248–9
Eigen, M., 130, 225–30, 238–42, 260
Einstein, A., 43, 245, 251, 256
Eldredge, N., 11, 131
Entropy, 253
ESS, 56, 74, 194–7, 200, 207–15
Essentialism, 8–10, 12

Eukaryotes, 172–3
Extinctions, 129–30, 185

Feedback, delayed, 226
Feldman, M.W., 62–7, 76–7, 117, 259
Fermat, P. de, 20
Ferrari, L., 20
Feynman, R., 29
Fincham, J., 127
Fischer, E., 103
Fisher, R.A., 8, 13, 14, 55, 99–101, 135, 143, 177, 203–4
Fox, S., 158
Franklin, R., 3, 5
Free Will, 100, 248–9

Galapagos Islands, 10
Galton, F., 13
Gambling, 20–1
Genetic Code, origin of, 241
Genetic Determinism, 82–3, 109–10
Gilbert, L., 214
Golschmidt, R., 35, 134–5, 146
Gorczynski, G.M., 160
Gould, S.J., 11, 21, 28, 88, 93–7, 125–30, 131, 134, 146, 148, 154, 259
Group Selection, 54, 106, 186
Gruber, H.E., 153
Guardian, the, 21, 24

Hadamard, J., 247
Haldane, J.B.S., 13, 14, 22, 26, 27, 32, 55, 88, 96, 110, 144, 187, 192, 247
Hallam, A., 130
Hamilton, W.D., 55, 101, 106–7, 197, 204
Hegel, G.W.F., 36, 37
Hennig, W., 133–4
Hermaphroditism, 102–3, 173–4
Hessen, B., 45

Hierarchies, 127
Himmelfarb, G., 11
Ho, M.W., 160, 161, 260
Hopeful Monster, 134–6, 153
Homosexuality, 39–40
Hull, D., 133
Huxley, J., 12, 22
Huxley, T.H., 12
Huygens, C., 20
Hybrid Vigour, 166–7
Hypercycles, 237–43

I Ching, 17–19, 41
Incest Avoidance, 58–9, 85
Inclusive Fitness, 107
Infanticide, 91
Isomorphism, 42, 232

Jones, J.S., 126
Jones, Peter, 27

Kimura, M., 161, 188
Kin Selection, 192–3
Kitcher, P., 86–92, 259
Kuhn, T., 46–8, 244
Kummer, J., 212–13

Lack, D, 105
Lakatos, I., 44
Lamarckism, 24, 31–2, 49, 180, 257
Lande, R., 136
Language, evolution of, 65–6
Laplace, P., 245, 251
Lawlor, L., 215
Leach, E., 76
Leibnitz, G.W., 6
Levins, R., 30–8, 258
Levi-Strauss, C., 58
Lewontin, R.C., 30–8, 69, 88, 207, 258
Linnaeus, 9, 12
Lions, 55–6
Lumsden, C.J., 57, 67–78, 80, 117, 259

Lyell, C., 12
Lysenko, T.D., 31–2, 38, 59, 159

Macaulay, T.B., 11
Malthus, T.R., 45, 114
Mammal-like Reptiles, 138–40
Marxism, 30–8, 44–5
Matsuno, K., 157,
Maxwell, J.C., 252
Mayr, E., 8–14, 35, 133, 146, 161, 258
Medawar, P.B. 24, 28, 96
Memes, 108
Mendel, G., 4, 10
Mendel's laws, 60, 119
Mendelians, 10–11, 47–8
Miller, Jonathan, 26
Mimicry, 159–60
Modern Synthesis, 10–11
Molecular Drive, 128
Molecular Genetics 36
Moment, B.G., 224
Monod, J., 50
Morgan, E., 40
Morgan, T.H., 12
Muller, H.J., 142
Mutation, 10–11, 20, 151, 166–7, 181
    rate of 182–4
Myth, 21, 39–50

Nash Equilibrium, 195–6
Needham, J., 17–18
Nelson, G.J., 158
Newton, I., 4, 6, 45, 251, 253

Oparin, A.I., 32
Origin of Life, 158, 237–42, 249
Origin of Species, 12

Packer, C., 198, 213
Pandas, 93
Pangloss's Theorem, 88
Parker, G.A., 210–11, 214

Parthenogenesis, 166, 168–71, 186
Pascal, B., 20
Pauling, L., 4, 7
Pearson, K., 38, 47–8, 59
Piaget, J., 75–6
Platnick, N., 158
Pollard, J.W., 160
Popper, K., 43–4, 46, 96, 244–9, 260
Price, G.R., 204–5
Prigogine, I., 250–7, 260
Prisoner's Dilemma, 196
Probability,
    propensity theory of, 247
Prokaryotes, 172
Punctuationism, 11, 24, 125–7, 131–6, 148–54

Queues, 198–200

Randomness, 15, 19
Raup, D., 142
Reciprocal Altruism, 197–8
Red Queen Hypothesis, 133, 168, 183, 185
Richerson, D., 114–21, 259
RNA, 237–42
Rorschach Test, 17
Rotifers, 168

Saunders, P.T., 159, 161, 260
Schuster, P., 238–42
Seilacher, A., 130
Sex,
    determination, 100, 104
    differences, 84–5, 173
    evolution of, 165–73
Sex Ratio, 98–101, 174–5, 203–4
Sexual Selection, 175–8
Shaw, G.B., 21, 40, 42
Simpson, G.G., 125, 151
Slijper, E.J., 135
Smith, Don, 39–40, 42
Snow, C.P., 225

Social Contract Game, 199–201
Species, 8–10, 126–7
   species selection, 128–9, 137–42, 146, 154, 168, 186
   variation in space, 152
Sperber, D., 15–17, 40–2
Stanley, S.M., 11, 131, 132, 149
Steele, E.J., 24, 25, 160
Stengers, I., 250–7, 260
Stern., 159
Structuralism, 16
Sumner, J.B., 153
Sunday Times, 24
Symbolism, 15–21, 40–2

Teilhard de Chardin, 94
Television, 22–8
Thermodynamics, 157, 257
Thomson, K.S., 135–6
Times, the, 24
Tools, evolution of, 66
Trivers, R.L., 197
Turing, A.M., 159, 222, 256

Van Valen, L., 133, 183
Vavilov, N.I., 159
Vrba, E., 129, 158, 161

Waddington, C.H., 75, 82, 127
Wake, D., 126
Wallace, A.R., 10, 13, 45, 94, 114
War of Attrition, 209–10
Watson, J.D., 3–7, 48, 258
Weismann A., 13, 189
Wells, H.G., 22
Werren, J., 101
Wicken, J.S., 157
Wiley, R.H., 198
Williams, G.C., 171, 185, 186
Williamson, P., 150
Wilson, E.O., 46, 53, 57, 67–78, 81–5, 89–91, 117, 259
Winkler, R., 225–30, 260
Wittgenstein, L., 230
Wolpert, L., 219, 224
Woolf, B., 36
Wright, S., 13, 14, 137, 143–6
Wynne-Edwards, V.C., 54–5, 106

X-ray crystallography, 3, 6

Yanomamo Indians, 70

Zhabotinsky's Reaction, 254–5